EXPERIMENTS IN GALLIUM ARSENIDE TECHNOLOGY

D. J. Branning and Dave Prochnow

TAB BOOKS Inc.
Blue Ridge Summit, PA

FIRST EDITION
FIRST PRINTING

Library of Congress Cataloging in Publication Data

Branning, D. J.
Experiments in gallium arsenide technology / by D.J. Branning and Dave Prochnow.
p. cm.
Bibliography: p.
Includes index.
ISBN 0-8306-9052-2 ISBN 0-8306-9352-1 (pbk.)
1. Gallium arsenide semiconductors. 2. Galllium arsenide semiconductors—Experiments. I. Prochnow, Dave. II. Title.
TK7871.15.G3B73 1988
621.3815′2—dc19 88-17065
CIP

TAB BOOKS Inc. offers software for sale. For information and a catalog, please contact TAB Software Department, Blue Ridge Summit, PA 17294-0850.

Questions regarding the content of this book should be addressed to:

Reader Inquiry Branch
TAB BOOKS Inc.
Blue Ridge Summit, PA 17294-0214

EXPERIMENTS IN GALLIUM ARSENIDE TECHNOLOGY

AD

Other Books in the Advanced Technology Series

Experiments with EPROMS
by Dave Prochnow
Edited by Lisa A. Doyle

Experiments in Artificial Neural Networks
by Ed Reitman
Edited by David Gauthier

Experiments in CMOS Technology
by Dave Prochnow and D. J. Branning
Edited by David Gauthier

To my dear and loving Kathy

Trademark List

The following trademarked products are mentioned in *Experiments in Gallium Arsenide Technology*.

Apple Computer, Inc.: Apple *II*e
Macintosh
Autodesk, Inc.: AutoCAD 2
Bishop Graphics, Inc: E-Z Circuit
Quik Circuit
Kepro Circuit Systems, Inc.: Kepro PCB
Ungar Electric: System 9000
Wahl Clipper Corporation: IsoTip 7800
IsoTip 7700
IsoTip 7240
Weller: EC2000
Wintek Corporation: HiWIRE
smARTWORK

Contents

List of Projects

Acknowledgments

Major contributions were made by seven manufacturers during the preparation of this book. Autodesk, Inc., Bishop Graphics, Inc., Borland International, GigaBit Logic, Inc., Heath Company, Kepro Circuit Systems, Inc., and Wintek Corporation each made generous hardware and/or software contributions that served as vital references for developing this book's text and projects.

Introduction

Never before has a semiconductor technology so totally captured both the heart and the mind of the electronics industry as has gallium arsenide (its chemical handle is GaAs). Declared as a supplement to the more conventional silicon semiconductor manufacturing ingredient, gallium arsenide is an exciting synthetic by-product from aluminum and copper mining (also lead refining) that sports seven significant advances over its silicon sister:

1. Gallium arsenide is faster than silicon. Electrons can be transported with GaAs up to 6 times faster than similar electrons in a silicon substrate.
2. Gallium arsenide is more *rad hardened* (radiation resistant) than silicon. Circuits designed with GaAs are more resistant to radiation influences than equivalent silicon circuits. This is an important feature in national defense weapons systems.
3. Gallium arsenide is more temperature tolerant than silicon. Major GaAs-based computer systems can be operated at higher temperatures thereby reducing the need for comprehensive cooling systems.
4. Gallium arsenide is less power-hungry than silicon. A given GaAs circuit requires less power than its silicon counterpart.
5. Gallium arsenide substrates emit photons. Similar silicon substrates are incapable of generating light.
6. Gallium arsenide has a higher degree of light absorption. This sensitivity to light is dramatically higher than the light reaction demonstrated by silicon.
7. Gallium arsenide circuitry is both electron and photon sensitive, simultaneously. This dual sensitivity enhances the data handling capabilities of the GaAs circuit.

All is not a bed of roses in the gallium arsenide garden, however. Hanging over the head of these seven remarkable attributes are three unfortunate detriments:

1. Gallium arsenide is expensive. Unlike ubiquitous silicon, GaAs must be synthesized from gallium bars and arsenide nuggets inside high-temperature furnaces.
2. Gallium arsenide is only capable of low-density circuit wafers. While a GaAs wafer can yield, at best, approximately 10,000-component density chips, silicon circuit designers can cram the same surface area with over one million such devices.
3. Gallium arsenide circuits are difficult to test. The blinding processing speeds of GaAs chips make the construction of suitable test equipment a major stumbling block in the manufacturing of defect-free GaAs chips.

Balancing these pluses and minuses is a circuit design nightmare that is currently haunting the microelectronic industry. On the flip side of the GaAs manufacturing coin is the optoelectronic industry. Manufacturing LEDs (*l*ight-emitting *d*iodes), IREDs (*i*nfrared-emitting *d*iodes), and laser diodes, this $500 million dollar branch of the GaAs tree has successfully overcome all of the potential production pitfalls and is currently flooding GaAs products into the lucrative commercial market. In turn, the increased revenues generated from these sales is prompting advances into the more profitable territory of GaAs circuit design.

Following the players in this rapidly expanding GaAs game can be difficult without a comprehensive "player's roster and playbook." This book is an attempt at providing this much needed reference text. Contained within its eleven chapters is a thorough summary of the products, manufacturers, and technologies that contribute to the gallium arsenide microelectronics and optoelectronics industries. Beginning with an elementary introductory chapter advancing the roots of the gallium arsenide technology, this book then documents ten separate GaAs products within ten distinct chapters. As an enhancement to the reader's studies, a consistent format is maintained throughout each of these chapters.

Leading each of these chapters is a brief introduction into the technology surrounding the featured gallium arsenide product. Supporting this introduction is a product reference section that highlights the specifications for a selected number of landmark GaAs products. Finally, concluding many of these chapters there are one or more advanced circuit construction projects that utilize the chapter's featured gallium arsenide product. These projects are incorporated into the text so the reader can apply each given GaAs technology in a practical situation.

Several thorough appendices conclude this examination of gallium arsenide technology. Project building suggestions, IC data sheets, manufacturer addresses, a section of suggested further reading, and a glossary make up this concluding comprehensive reference portion.

During your study of this fascinating field, remember that the microelectronics industry is currently facing a technological rebirth, and the name of that technology is *gallium arsenide*.

1

Gallium Arsenide Technology

Light is the physical excitation of discrete photons along a wide range of spectral wavelengths. This greatly generalized definition encompasses an enormous variety of different "lights." Infrared, ultraviolet, and visible light are all members of this excitable light spectrum. This wave theory of light has not always been so easily defined, however.

Early studies in the physics of light began with observations of the brightest source of natural light—the sun. Based on these initial scientific inquiries, several elementary light theories were established. The first of these theories dealt with the brightness of the light generated by the sun. In general, it was determined that the intensity of sunlight falling on the earth's surface was variable due to atmospheric conditions. The unit of measurement used in this theory was the *footcandle*. One footcandle equalled the amount of light produced by one single candle burning at a distance of one foot. With this unit in hand, the greatest intensity of sunlight on the earth's surface was measured at 10,000 footcandles.

In addition to its extreme brightness, the natural light from the sun was also postulated to travel in waves. These waves behave exactly like their counterparts that are found in water. Unfortunately, this theory wasn't adopted for several centuries. Instead, the ideas of Isaac Newton predominated the scientific community. Briefly, Newton stated that light consisted of particles or corpuscles and that these corpuscles moved in straight lines. Another light theory that rivaled Newton's was that of Christian Huygens who stated that light moved in waves. Both of these theories—Newton's corpuscular theory of light movement and Huygens' wave theory of light movement—were accepted as truthful explanations for the physics of light in the late 1700's.

In 1801, Thomas Young designed an experiment that proved that Huygens' theory was correct. This series of experiments and counter-experiments became known as Young's Interference Experiment.

YOUNG'S INTERFERENCE EXPERIMENT

Purpose: To test the wave theory of light through a series of interference experiments.

Materials: dishwashing soap
glass dish
small wire loop (1 inch in diameter)
clear nail polish
sheet of white cardstock (3 inches by 5 inches)

Procedure:

✤ Place several drops of dishwashing soap into a glass dish filled with water.
✤ Dip the wire loop into this solution. Note the film formed across the wire loop.
✤ Make several large soap bubbles by swishing the soap/water solution around. Observe the color changes inside each of the bubbles.

OR,

✤ Take the clean glass dish and fill it with water.
✤ Add two drops of nail polish to the water. This produces a nail polish film on the bottom of the glass dish.
✤ Punch three, round, clean-edged ¾-inch holes across one end of the white cardstock.
✤ Dip the holey end of the card into the solution and slowly slide it through the nail polish film on the bottom of the glass dish so that films form across the holes.
✤ Observe the color changes through the film that has formed on the cardstock holes by holding it up to a light, and compare them to the appearance of the same film as you hold the card down so light *reflects* off the surface of the cardstock.

Results: What conclusion can you draw from the numerous color changes that each of these experiments generated? Thomas Young was able to use the conclusions from these experiments to prove Huygens' wave theory of light.

ELECTROMAGNETIC SPECTRUM

Optical light is only one small portion of a broad range of wave radiations. A collective reference for these radiations is the *electromagnetic spectrum*.

Building on the wave theory of Huygens, James Clerk Maxwell developed the electromagnetic theory of light in 1860. Maxwell's theory stated that light waves were of an electrical nature. This theory defined the numerous wavelengths of light where visible light shares the electromagnetic spectrum with various other waveforms. In terms of diversity, the electromagnetic spectrum spans from low-frequency audio waves (e.g., 300 Hz or 1,000,000,000 m) to ultrahigh-frequency gamma rays (e.g., .00000000000001 m). Due to this enormous disparity in the measurement of electromagnetic waves, several different units of measurement are frequently used in the expression of a wave's length.

For example, a wave's frequency can be determined by the equation:

$$f = v/W$$

where,

f = frequency in hertz
v = 186,000 miles/second
W = wavelength in meters

Other units of wavelength (symbol is λ) measurement include: microns (10^{-6} meters; μ), nanometers (10^{-9} meters; nm), and angstroms (10^{-10} meters; Å). Figure 1-1 illustrates the relationship of these various units of measurement with a significant portion of the electromagnetic spectrum.

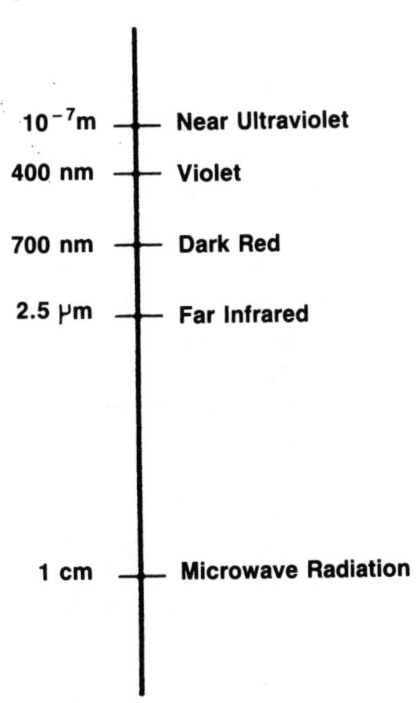

Fig. 1-1. A comparison of a selected portion of the electromagnetic spectrum and corresponding approximate wavelengths.

INFRARED WAVELENGTHS

Occupying the longer wavelengths of the optical spectrum is the infrared (IR) spectrum. There are three distinct groupings within the IR spectrum: the near infrared (NIR) has a wavelength range between 770 nm and 1.5 microns, the intermediate infrared (IIR) ranges from 1.5 microns to 5.6 microns, and the longest waves are the far infrared (FIR)—between 5.6 microns and 1,000 microns. The longer wavelengths of the far infrared spectral portion mark the upper extremes of the microwave wavelengths.

Radiation absorption and transmission are two important properties of infrared. In order to appreciate the application of IR wavelengths, a complete understanding of the interconnection between the absorption and transmission of various objects is necessary. There are several key physical relationships that simplify this understanding.

Before these relationships and their resultant formulas can be discussed, however, an absorption/transmission definition must be made. In IR radiation, this definition is the *black body*. A black body is a hypothetical unit which is capable of absorbing all incident radiation. Furthermore, the radiation emission of a black body is an ideal for a given temperature. Therefore, the black body is the definition of optimal IR absorption and emission.

Accompanying this definition is a unique unit of IR absorption and emission measurement. Known as *emissivity*, this unit forms a ratio of the radiant energy emitted by an object to that emitted by a black body at the same temperature. In other words, emissivity is the relative power of an object to emit IR radiation. Based on this relationship, a black body has an emissivity of 1.00. Operating from this ideal condition, the emissivity of any object can be determined.

Three functions govern the establishment of an object's emissivity. Beginning with Planck's black body definition:

$$w(\lambda,T) = C^1 / \lambda^5({}_eC_2 / \lambda T - 1) \text{ watt cm}^{-2}\mu^{-1}$$

In other words, the black body power radiation per wavelength at a unit temperature.

Next is Wien's Displacement that states that peak wavelength radiation is inversely proportional to the absolute temperature of the object. This law expressed in formula states:

$$W = k / T$$

where,

W = wavelength in microns
k = a black body constant of 2900
T = temperature in degrees Kelvin

Finally, the Stefan-Blotzmann function provides that the radiation intensity of an object is directly proportional to the product of the object's emissivity and the fourth power of the object's absolute temperature.

$$Wd\lambda = k * e * T^4$$

where,

$Wd\lambda$ = intensity
k = constant of $5.679 * 10^{12}$
e = object emissivity
T^4 = temperature in degrees Kelvin

Objects that approach the emissivity of a black body can be found in cavity openings. An ideal example of this opening is the exhaust port of a turbojet engine's tail pipe. Representatives of these objects exhibit wavelength radiation peaks of 3.3 microns. This is an important wavelength for optimal range performance within passive infrared detection systems. Four assumptions that are clinically made by these detection systems are:

1. Both the detection system's response time and the infrared radiation source are matched in their transmission rates. A *chopping reticle* is a common means for generating this detection system carrier frequency.
2. The sensitivity of the detection system is limited by its current noise.
3. The detection system exhibits a minimal signal-to-noise ratio of 4.
4. The IR source falls completely within the detection system's sensor.

Basically, these four assumptions can be categorized within two broad characteristics: transmission characteristics and detector characteristics. In the first case, transmission characteristics deal with the attenuation of the IR signal. Both atmospheric and optical absorption can result in the attenuation of these IR wavelengths.

Particulate wavelength scattering and *molecular absorption* are two of the primary problems associated with atmospheric absorption. In the former transmission characteristic, suspended particles in the atmosphere contribute to the deflection of IR wavelengths. Fortunately, the uniform attenuation produced through particulate scattering can generally be compensated for through broadband signal amplification. This is not true with molecular absorption. This atmospheric absorption problem is, to a large extent, limited to the molecular resonance of water and carbon dioxide molecules.

The resonance of these two molecules occurs in the infrared frequency range. Furthermore, the molecular resonance of water and carbon dioxide molecules establishes an absorption-enhancing electrical oscillation. The net result from this additional charge vibration is the attenuation of IR wavelengths.

Supplementing atmospheric absorption in the attenuation of IR wavelengths is optical absorption. The losses suffered through the employment of optics in IR detector systems can cloud the benefits that are derived by their use. For example, optics are vital for the optimal gathering and focusing of IR wavelengths on the detector's surface. By the same token, internal reflections, transmission degradation, and physical breakdown are all features of optical systems that serve to attenuate IR wavelengths.

Countering the wavelength attenuation found in atmospheric and optical absorption is the performance parameters of the detection system. These detector characteristics center on the noise of the system and the responsiveness of the detector substratum.

Noise is a performance-degrading by-product of electrical current flow. The deleterious nature of circuit noise is radically reduced through IR detector system shielding. Current flow can also contribute to the formation of another noise—thermal noise. Unlike circuit noise, thermal noise is far more detrimental in an IR detector system. Based on the reduced IR signals that are exposed to an IR detector, internal thermal noise produces errant readings resulting in the failure of the detection system. The simplest solution to the reduction of thermal noise is through the elimination of unfiltered power supplies.

Understanding the inherent noise of an IR detector system is directly linked to an expression of the sensitivity of the detector. A common method used for expressing this noise-to-sensitivity relationship is through a signal-to-noise ratio (SNR). In elementary terms, a signal-to-noise ratio of 1 represents a detector signal response that is equal to its system noise. Therefore, the higher the SNR, the greater the degree of spectral sensitivity over a reference wavelength.

The final performance parameter of an IR detector system is the physical responsivity of the detector. This characteristic is based on the detector's material or substratum. In order to determine the responsivity of a detector, use the formula:

$$R = O/I$$

where,

R = responsivity in volts per watt
O = output signal; reverse current in microamperes
I = input radiation intensity in milliwatts per square centimeter

A factor affecting the responsivity of a detector is the spectral response wavelength. Simply put, this is the wavelength where the detector sensitivity is altered. In other words, the spectral response wavelength is the wavelength where the responsivity of the detector is halved. This factor is atmospheric-dependent and therefore requires a strict adherence to performance controls.

One popular misconception of infrared wavelengths centers on their confusion with laser (*l*ight *a*mplification by *s*timulated *e*mission of *r*adiation) beams. Generally speaking, lasers can be formed from infrared wavelengths, but not all lasers use infrared wavelengths. This concept is explained more fully in Chapter 8.

THE PHYSICS OF LIGHT

Defining light, based solely on Huygens' wave theory, is inadequate when dealing with semiconductor-generated light sources. An additional physical understanding of light is necessary. This supplement is the basic quantum theory of light.

Essentially, the basic quantum theory of light builds on the wave theory by adding the concept of light's particulate composition known as *photons* (this is an adaptation of Newton's corpuscular light theory). In this regard, photons are uncharged particles with a wavelength-established energy level. In other words, each photon's energy level is fixed through its wavelength. Therefore, based on their shorter wavelength (or higher frequency), ultraviolet photons have a greater energy level than infrared photons. This fact is easily demonstrated by the painful sunburn that can be quickly acquired during unprotected summer exposure to the sun.

The propagation and control of both light's waves and its photons mandates the application of optics. In this context, the term optics isn't a strict reference to conventional glass lenses; rather, optics concerns the impact of light rays with a defined surface.

The first of these light ray definitions is the law of reflection. Basically, a light ray reflection occurs whenever light interfaces with two dissimilar media. For example, think of a light beam striking the surface of a silvered mirror. In this example, the two dissimilar media would be the air through which the light beam is traveling and the silvered mirror that the light ray is striking. This law is known as Snell's Law of Reflection.

Snell's Law states that the angle of reflection of the light ray equals to the angle of incidence. In supporting the definition of this law, both of these angles are measured from a line that is perpendicular to the reflecting surface at the point of reflection. This line is called the *normal*. Furthermore, all three of these features—the angle of reflection, the angle of incidence, and the normal—all lie in the same plane (see Fig. 1-2).

This initial statement of Snell's Law deals with light rays striking an opaque surface. Another facet of Snell's Law also provides for a similar ray of light striking a translucent surface. In this application, a portion of the light ray will pass through or be refracted by the interface surface. The definition of this refraction is provided in Snell's Law of Refraction.

Contrary to the reflection of a light ray, a refracted light ray does not pass through the translucent surface in a fixed angle of incidence. Instead the density of the interface surface causes the speed of the light ray to be slightly reduced. The reduction in the light ray also reduces the propagation of the light through the interface surface. In turn, this results in the bending of the light ray.

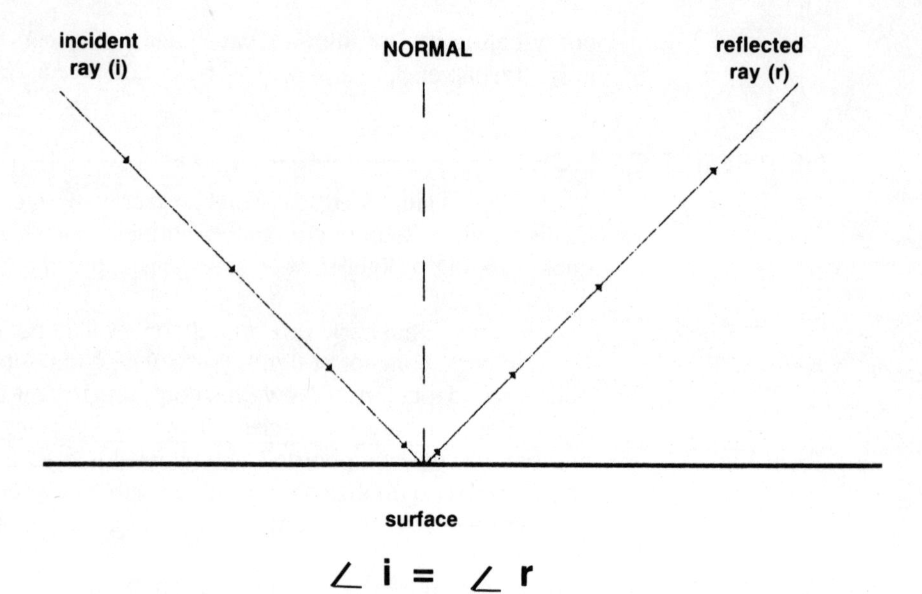

Fig. 1-2. The relationship between the angle of reflection and the angle of incidence.

Expressing the refraction of a light ray through a translucent surface is performed by a refractive index. The refractive index is the ratio of the speed of light in a vacuum to the speed of the light ray traveling through the translucent interface surface. Expressed in formula terms,

$$n = L / V$$

where,

n = refractive index
L = speed of light in a vacuum (186,000 miles per second)
V = speed of light ray through translucent interface surface

Once the refractive index of a material is known, then the degree of refraction can be determined with Snell's Law of Refraction. This law is the function of the relative difference in the refractive index of the incident medium (for example, the air), the refractive index of the refractive medium (for example, a pane of glass), the angle of refraction, and the angle of incidence (see Fig. 1-3). Expressed as a formula,

$$\sin 0_i / \sin 0_R = n_R / n_i$$

where,

0_i = angle of incidence
0_R = angle of refraction
n_R = refractive index of the refractive medium
n_i = refractive index of the incident medium

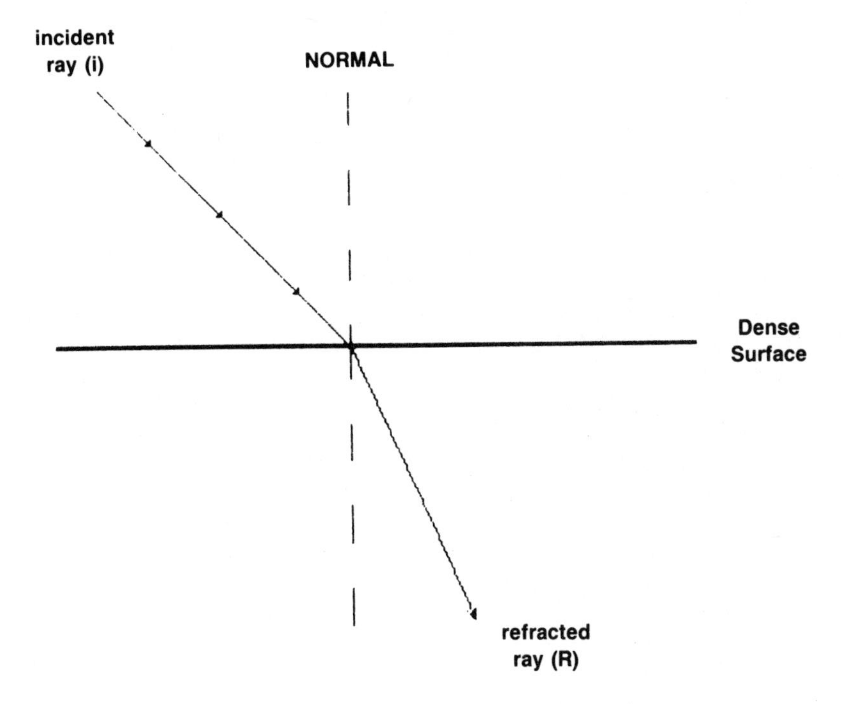

Refractive Index = Sin ∠ i / Sin ∠ R

Fig. 1-3. The relationship between the angle of refraction and the angle of incidence.

Once the light ray has passed through the refractive medium, it exits back into the incident medium (or, a different incident medium). This exit angle is $0_i'$ or arcsin (Sin 0_R). The degree of this exit angle is a function of the density of the refractive medium versus the incident medium (see Fig. 1-4).

When dealing with refractive surfaces, it is possible that the angle of incidence can be moved to an extreme angle that results in an angle of refraction that is parallel to the refractive surface. This angle is known as the critical angle. Increasing a light ray's angle of incidence beyond the critical angle can result in a refractive surface reflecting the light ray in an angle of reflection that is equal to the angle of incidence. This reflection is known as *total internal reflection*. Chapter 9 provides a more in-depth examination of this action with reference to optical fibers.

Three important alterations can be made to light rays interfacing with optical surfaces. The first of these alterations is the loss of energy through *Fresnel reflection*. Fresnel reflection is derived from the refractive index difference of the refractive medium and the incident medium and the angle of incidence (see Fig. 1-5). Another light ray alteration comes in the form of

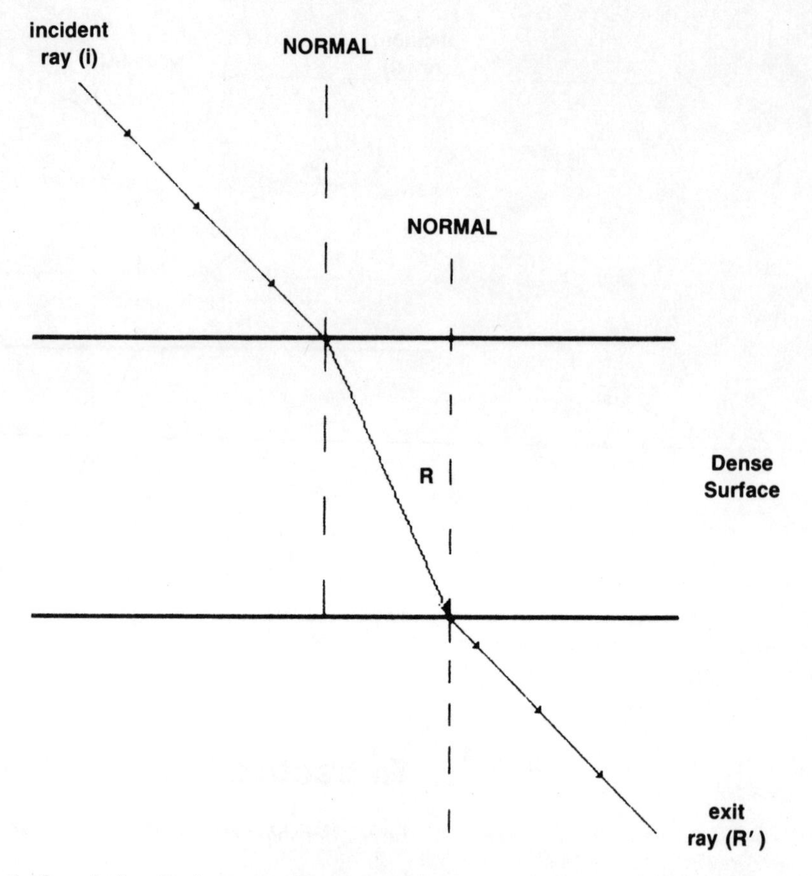

Fig. 1-4. The relationship between the exit angle of a refracted light ray and the angle of incidence.

refractive medium absorption. Colored filters are excellent examples of translucent materials that absorb light rays. For example, a red filter appears to "color" light red, while, in fact, the red filter has actually absorbed all of the other colors of light (notably, green and blue) and allows only red light to pass through. This absorption doesn't actually destroy the other colors of light, rather it converts their energy into another form, such as heat.

The final alteration of light deals with the spreading or diffusion of light rays. In contrast to the smooth surface of a reflective or refractive optic, a diffusive optic is marked by a rough surface. This rough surface scatters or reflects light rays in an assortment of different angles. Therefore, the intensity of light passing through a diffused optic is equal from all viewing angles. Many LED light bars use this optic for optimal indication.

Measurement in the physics of light is conducted through *radiometry* and *photometry*. Each of these disciplines provides a degree of overlap in their

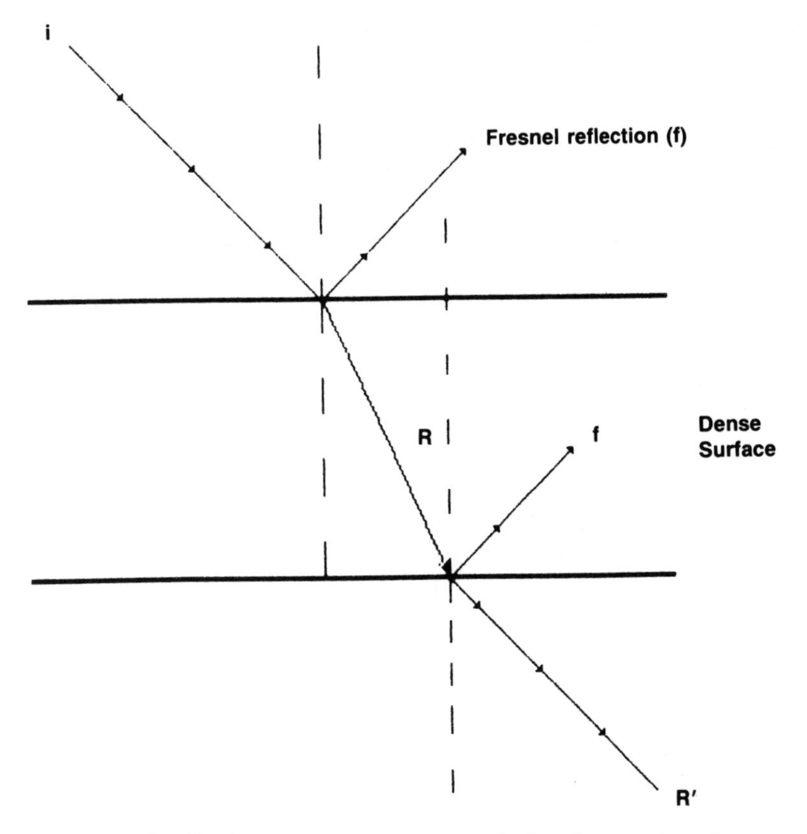

Fig. 1-5. Fresnel reflection can occur on any optical surface—external as well as internal.

examination of light. Radiometry measures all radiant energy in the electromagnetic spectrum. Photometry, on the other hand, measures radiant wavelengths inside the visible optic spectrum. Regardless of the wavelengths measured by these two disciplines, they both adhere to a standard set of criteria.

flux—the rate at which energy passes through a medium; energy per time unit.
energy—the energy level at the electromagnetic source.
incidence—the energy rate exciting a surface; flux per area unit.
exitance—the energy rate exiting a surface; flux per area unit.
intensity—the energy rate of a solid angle radiating from a source; flux per solid angle unit.
solid angle—the ratio of a sphere's surface area to the square of the sphere's radius; one sphere unit.

THE SOURCES OF LIGHT

Sources of light radiation can be either natural, like the sun, or artificial, such as from semiconductor materials. Foundation studies in the basic quantum theory of light were conducted almost exclusively with natural light. These same results can now be tested and confirmed with artificial light.

In the category of artificial optic radiation, there are currently three major methods of light generation: *incandescent*, *illuminescent*, and *semiconductor* (although semiconductor radiation is a form of illuminescent radiation). Incandescent radiation was formerly introduced in 1879 by Thomas A. Edison. Very little has changed between this initial incandescent offering and today's conventional household light bulb.

An incandescent light source generates its optic radiation through the heating of a metal filament housed in a controlled gas atmosphere. As this filament is heated, it incandescences and forms both light and heat radiation. The type of spectral wavelengths that can be generated in incandescent lamps is limited to the filament composition and temperature and the absorption of the surrounding glass sphere.

The second major form of artificial light radiation is the illuminescent lamp. A typical example of an illuminescent lamp is the gas-discharge or neon lamp. These small glass indicators consist of two electrodes housed inside a glass vial. The air inside this glass enclosure has been removed and replaced with either neon or argon gas. These gases produce a unique color display with the neon generating longer wavelengths of red light and argon producing shorter blue wavelengths.

Gas-discharge lamps glow when the electrical potential between the two electrodes is raised above the breakdown voltage level for the gas. This causes the gas to ionize and the lamp to glow. Furthermore, the electrodes of the gas-discharge lamp can assume any polarity. Therefore, either ac or dc voltages can be applied to the lamp with equal success.

The final source of artificial light is semiconductor radiation. The remainder of the text is devoted to the study of semiconductor radiation technology.

GaAs TECHNOLOGY

The employment of gallium arsenide (GaAs) in semiconductor technology is a quest for the highest possible speed of processing at the IC level. This speed factor is possible because GaAs electrons are more mobile than their silicon counterparts. Basically, the field of GaAs technology can be divided into that of optoelectronics and digital ICs. Both of these approaches are thoroughly discussed in this book.

Gallium arsenide optoelectronics include a large family of LEDs, displays, optocouplers, and laser diodes. Each of these singular components share a similar fabrication technology along with an equally common product—light.

Light in a GaAs optoelectronic device, such as an LED, is produced following the application of a forward-bias current (see Fig. 1-6). This current application generates electrons for injection into the n-type semiconductor medium. A result from this injection is the evacuation of electrons from the p-type medium. This electron withdrawal forms holes in the p-type medium.

A recombination of these electrons and holes occurs at the p/n junction. This action generates photon energy. Subsequently, optics can be used on these photons to propagate and control the radiated light rays.

Most LEDs use various gallium compounds in the doping of the n-type medium. Each of these compounds offers a different form of photon production.

gallium arsenide—GaAs; infrared radiation.
gallium arsenide phosphide—GaAsP; long-wavelength red radiation.
gallium phosphide—GaP; shorter wavelength green radiation.
gallium aluminum arsenide—GaAlAs; long wavelength red radiation.

Of these four doping compounds, GaAs is the most efficient wavelength radiation source. By producing photons in the infrared spectrum, GaAs LEDs generate a brighter, more optimal output than the shorter wavelength red and green LEDs. This prominent photon activity ideally lends IR LEDs to communication applications.

The use of GaAs technology in digital ICs is significantly different from that used in optoelectronics. In a digital GaAs IC, photon production is not the desired product; instead, increased electron transfer speed is the goal. This electron acceleration is commonly on the order of six times faster than any speed found in a comparable silicon-based IC. Several other improvements

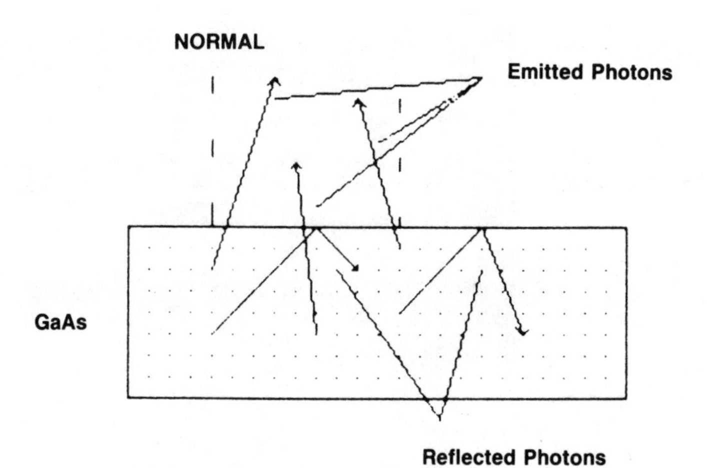

Fig. 1-6. Emission angles from GaAs-generated photons.

found in GaAs IC technology strengthen the position of this fabrication technique over the previously discussed semiconductor technologies.

- Enhanced substrate insulation from destructive capacitances.
- Low power dissipation.
- High data rate transference.
- Wide tolerance range to outside temperature influences.
- Profound hardness to outside radiation levels.
- Capability to match I/O performance of any existing logic family technology.

Supplementing this impressive list of performance features, gallium arsenide digital ICs can be built from a number of different elemental circuits: D-MESFET (*d*epletion-mode *m*etal-semiconductor *f*ield-effect transistor), E-MESFET (enhancement-mode MESFET), and HEMT (*h*igh-electron-*m*obility *t*ransistor). Of these three basic circuits, D-MESFET is the most commonly encountered.

An example GaAs IC fabricated from the D-MESFET process technology is the GigaBit Logic Quad 4:1 or Dual 8:1 Data Multiplexer (see Fig. 1-7). GigaBit Logic's 10G046 is a general-purpose multiplexer (mux) with bus, data and address, and fan-in data channeling capabilities. This PicoLogic family member can have its 8:1 outputs ORed together for expansion to a 16:1 mux.

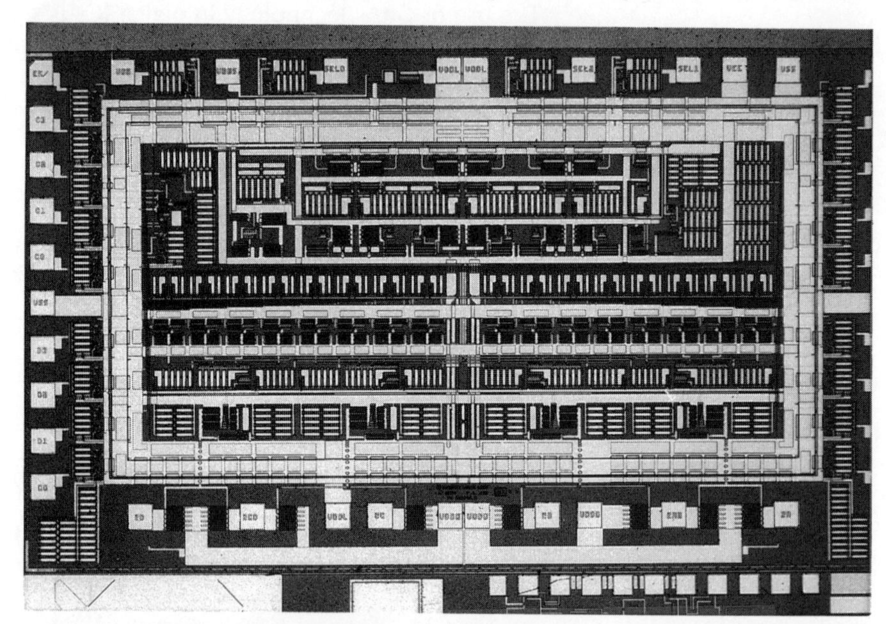

Fig. 1-7. A D-MESFET process technology digital GaAs IC—GigaBit Logic's 10G046 Quad 4:1 or Dual 8:1 Data Multiplexer. This is a top view of the internal logic architecture. Photograph courtesy of GigaBit Logic, Inc.

A D-MESFET has a doped n-type implant region that is 100 to 200 nm deep. The source and drain contacts are placed 3 to 4 microns apart. Control of the D-MESFET is manipulated through gate voltage switching between the source and the drain. Essentially, the operation of this D-MESFET circuit follows that of a n-channel-junction FET NMOS circuit. Two points that separate the D-MESFET from the NMOS circuit, however, are the high transconductance coupled with the low input capacitance of the GaAs-based circuit. This results in a high-gain-bandwidth D-MESFET circuit that can post extremely fast switching speeds in the neighborhood of 50 picoseconds.

DIGITAL COUNTER

Second to the LED, the most visible (pun intended) GaAs technology component is the common-anode seven-segment LED display. This versatile 14-pin dual-inline package (DIP) is able to display all ten members of the base-10 number system based on the selective lighting of two or more of its seven segments.

In this circuit, the dual 7447 ICs (U1 and U2) receive a 4-bit code sequence from the decade counter ICs (U3 and U4). This code pattern produces a high output on a combination of the LED display's pins. The speed of displaying, advancing, and displaying the numbers on the displays is governed by the clock speed of the units decade counter IC.

Construction Notes

The construction of the Digital Counter is extremely simple with the wiring schematic and parts list provided in Fig. 1-8. Only three components will need to be added to this schematic for completing the counter circuit. First, a clock circuit must be attached to pin 8 of the units display's decade counter IC (U3 74196). This clock circuit can be easily designed from a 555 timer IC.

Second, a start/stop switch and a reset switch are necessary for manipulating the progression of the display. These switches can be placed in two locations in the Digital Counter circuit: the start/stop switch between the clock and pin 8 of U3 and the reset switch between the power supply and the power bus of the circuit.

One final component that will enhance the performance of the Digital Counter is a clock rate control. This control is usually a potentiometer in the clock circuit. By varying this control, the Digital Counter can be made to advance at any speed. For example, 0 to 99 seconds or 0 to 99 minutes can be selectively counted on the twin LED displays.

BURGLAR ALARM

Another readily accessible GaAs optoelectronic component is the infrared LED or IRED (*i*nfrared-emitting *d*iode). When combined with a suitable detector circuit, the IRED can be the source of a powerful burglar alarm.

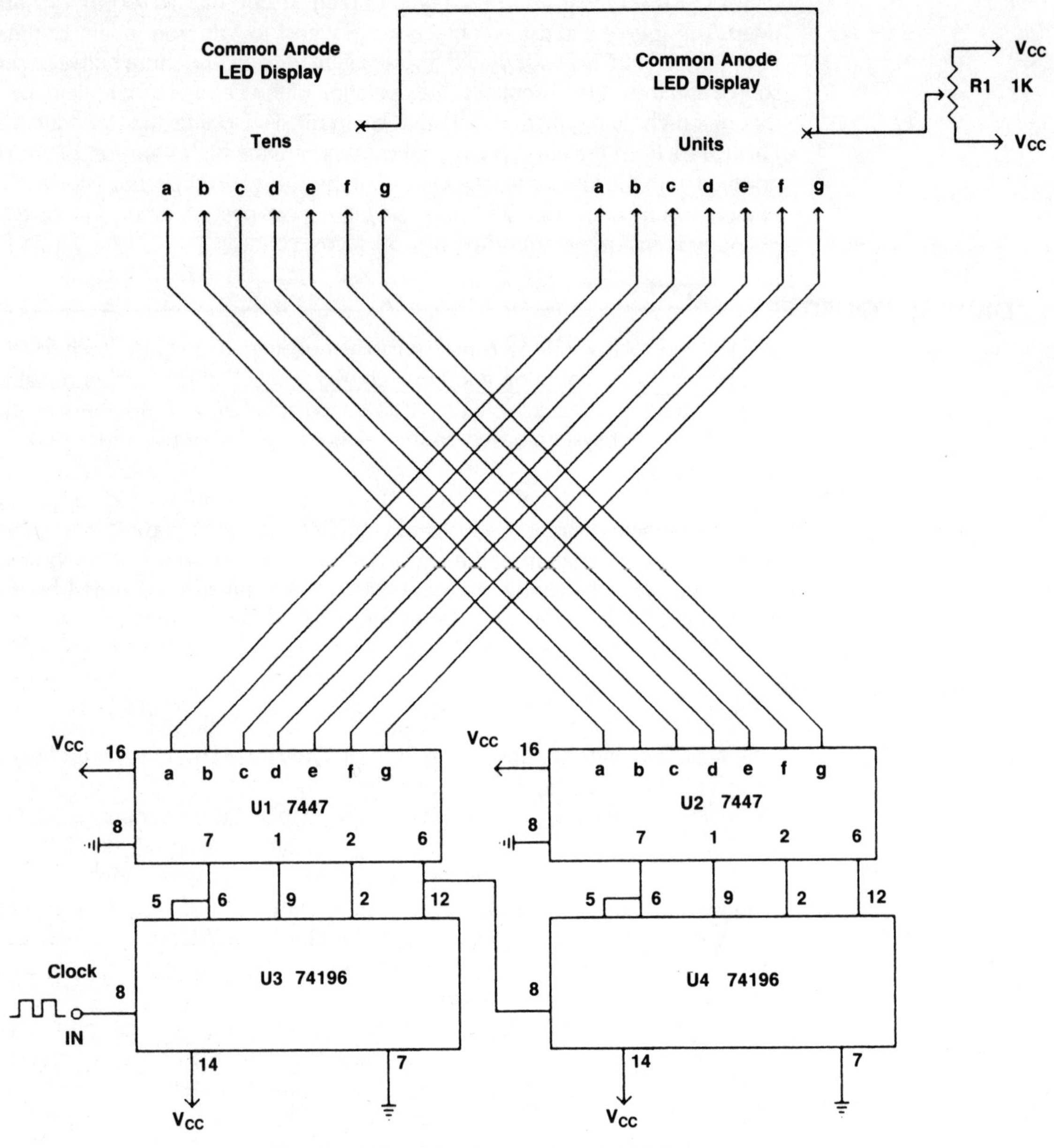

Fig. 1-8. Schematic diagram for Digital Counter.

Virtually, any IRED can be used in this project. The specified IRED can be substituted with any of the IR radiation lamps that are listed in Chapter 3 (for example, a TIL32 diode). If an IRED, different from the one specified, is used in this project, make sure that the phototransistor is spectrally matched to the radiation output of the diode, for example, a TIL78 phototransistor is matched to the output of the TIL32 diode.

Gallium arsenide technology is also used in components whose IR emission is invisible to the observer. These devices are known as optocouplers. Most optocouplers are used as connections between two dissimilar circuits with the switching, relay, feedback, coupling, or phase response effected optoelectronically inside an opaque DIP carrier. The integrated circuit U1 (4N26) in this project is an example of an optocoupler. In essence, this optocoupler consists of an internal IRED and an optically-coupled phototransistor. During operation, the detection signal from the exterior IRED/phototransistor pair is transferred through the optocoupler's internal IRED/phototransistor pair to an alarm output. This link provides an adequate demonstration of the linking abilities of the optocoupler, as well as the electrical isolation afforded by this optoelectronic connection.

Construction Notes

An assembly schematic for the Burglar Alarm is illustrated in Fig. 1-9. This figure also lists all of the values for this project's components. As previously mentioned, the designated IRED and phototransistor can be replaced with any other readily accessible matched optoelectronic pair. If a different IR emitter/detector pair is substituted for D2/Q1, however, the values for the support components might need to be adjusted. This adjustment would include resistors R5 and R6 and capacitor C1. Consult the component listings in Chapter 3 for determining the values required by this substituted IR detection system.

During operation, the Burglar Alarm will sound (provided a suitable audio output has been attached to pin 5 of optocoupler U1) whenever an object passes between D2 and Q1. The sensitivity of this response can be adjusted with potentiometer R3. Two other factors also affect the sensitivity of this IR detection system: distance and radiant intensity.

This first limitation refers to the distance between the IRED and the phototransistor. As illustrated, a distance of under 1 foot provides excellent sensitivity under normal to subdued room lighting conditions. In other words, bright, direct sunlight will grossly retard the sensitivity of Q1. By using glass lenses, however, the distance between D2 and Q1 can be safely increased to several feet with little degradation in sensitivity. Once again, this statement holds true only for normal room lighting conditions.

The second sensitivity-limiting factor deals with the radiant intensity or output intensity of the IRED. For example, the brighter the IR output from D2, the greater the distance can be between Q1 and D2. These different output levels can be obtained be varying the values for resistor R6. There are practical limits to the degree of radiant intensity that can be obtained from any given IRED, however. The limits for this output are given in the component listings in Chapter 3.

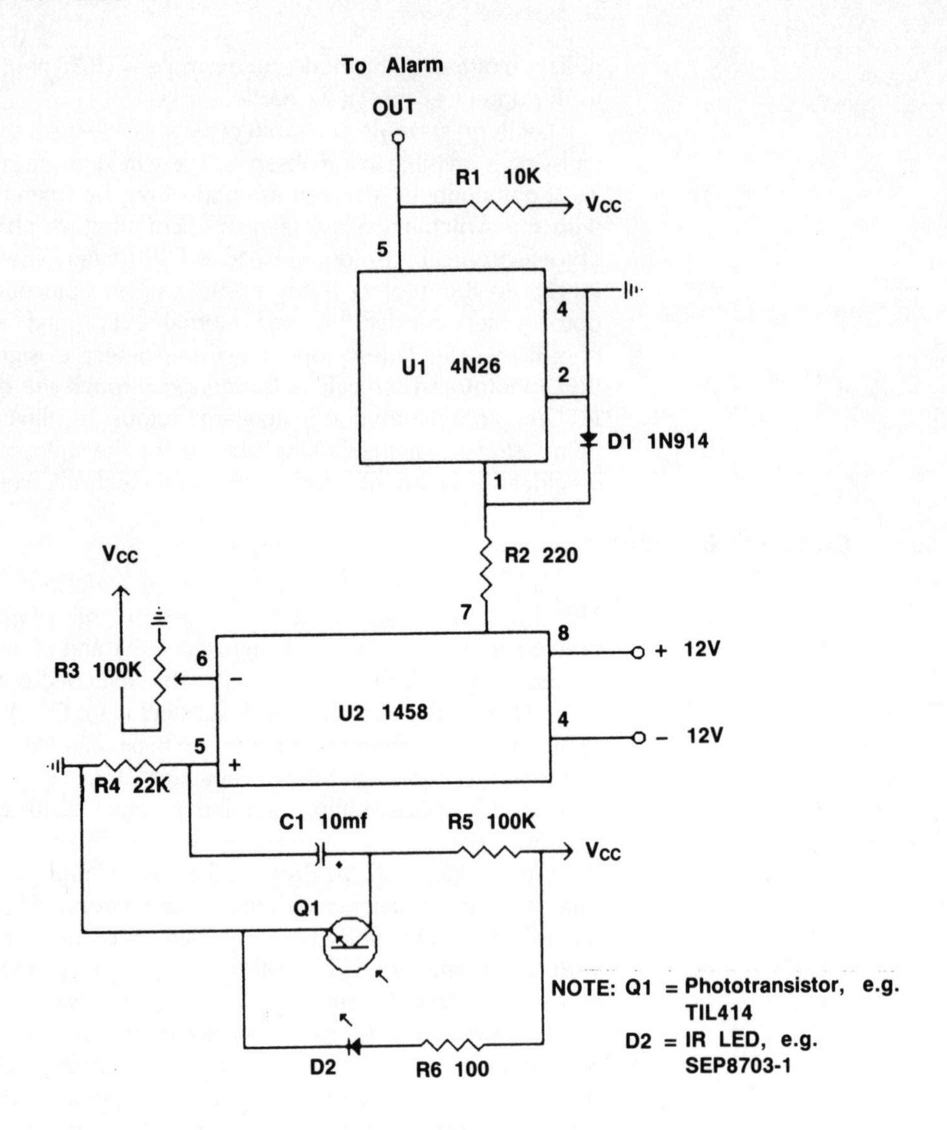

Fig. 1-9. Schematic diagram for Burglar Alarm.

Advanced Study

One final application for the Burglar Alarm is the result of combining the Digital Counter circuit with the optocoupler's output. This combination would create a project that can display the number of "trips" that have been registered between D2 and Q1. A possible arrangement for this project is given in Fig. 1-10. Based on this schematic diagram, a PCB (Printed Circuit Board) template and a parts layout are suggested in Figures 1-11 and 1-12, respectively.

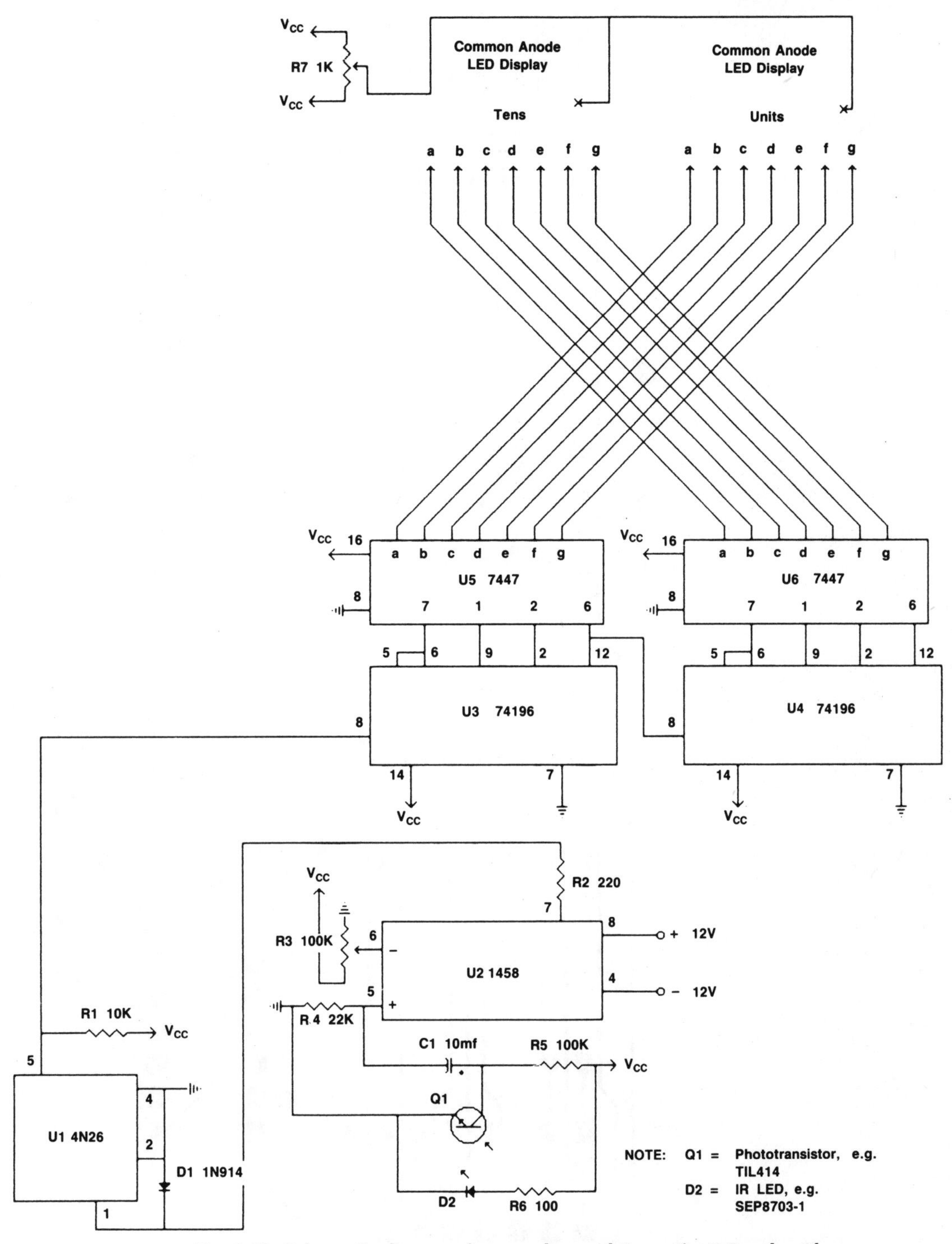

Fig. 1-10. Schematic diagram for an advanced "counting" Burglar Alarm.

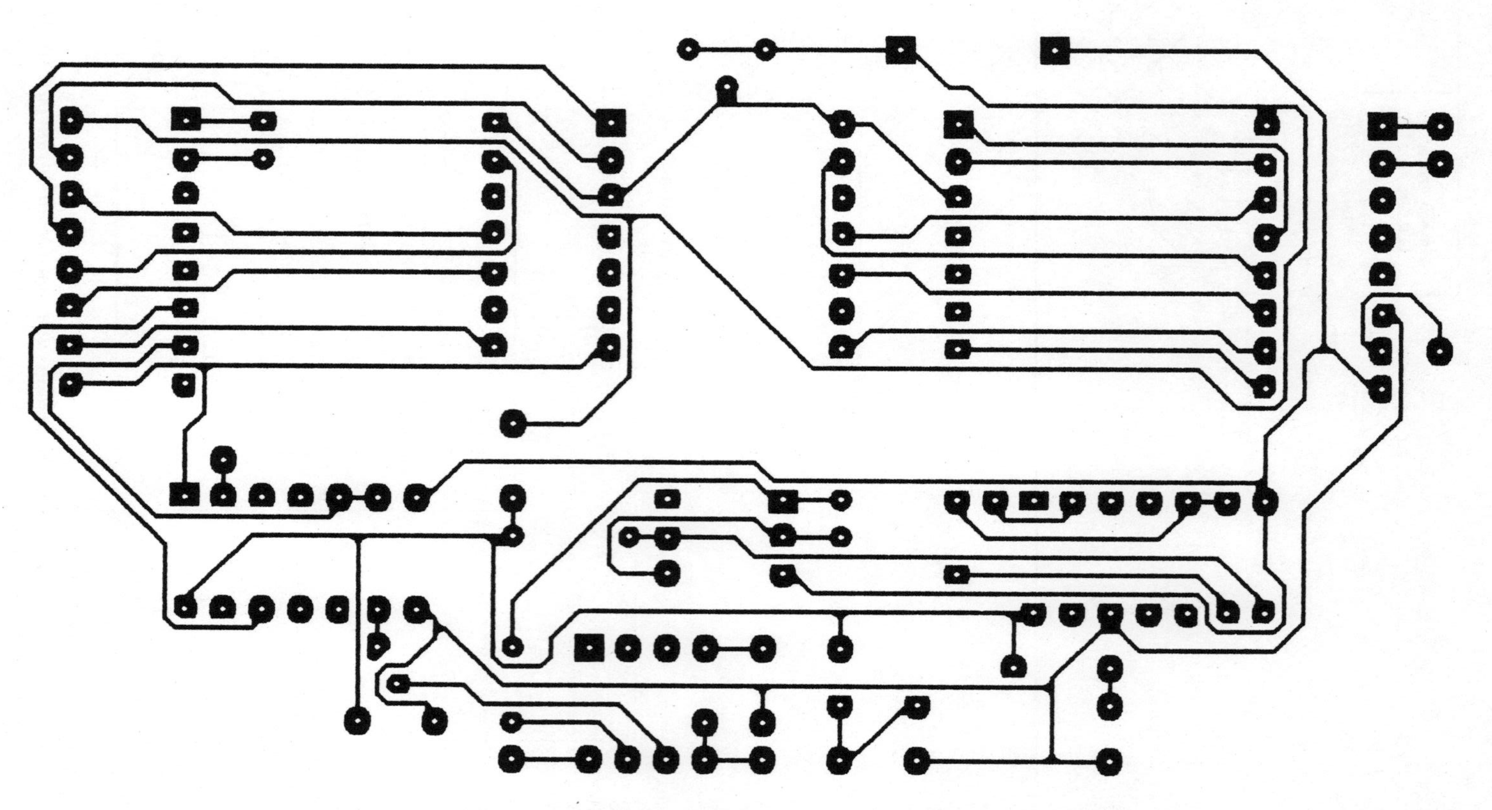

Fig. 1-11. Solder side for the "counting" Burglar Alarm template, shown at 2X size.

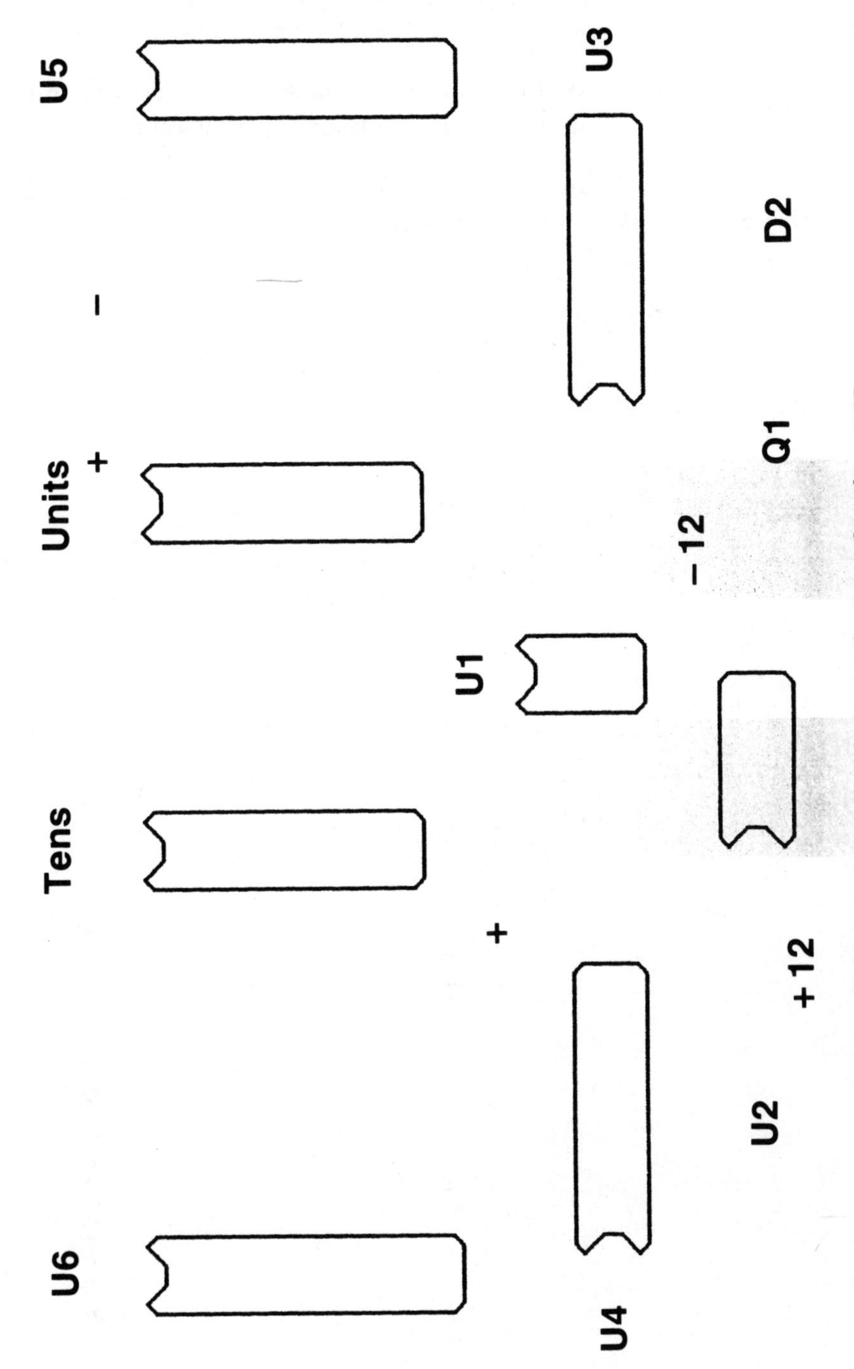

Fig. 1-12. Parts layout for the "counting" Burglar Alarm PCB.

In building this "combined" project, three separate GaAs technologies and properties can be demonstrated.

- The sensitivity of the IRED when used in an IR detection system.
- The electrical isolation of the optocoupler when used as an in-circuit coupling device.
- The versatility of the seven-segment LED display when used as a system status readout device.

2

GaAs FETs

Field-effect transistors (FETs) perform two distinct functions as applied to gallium arsenide technology. The first is in the role of discrete, small-signal, dual-gate n-channel transistors. In this configuration, GaAs FETs are useful as low-noise, high-gain signal receiver amplifiers.

The second utilization of GaAs FETs is in the fabrication of D-mode MESFET IC devices. Broad manufacturing tolerances and extremely high-speed switching capabilities make these GaAs D-MESFETs more attractive than their closest silicon JFET (Junction FET) rival.

Beginning with the discrete FET, its basic construction and operation parallels that of a typical silicon JFET. These JFETs are formed on a lightly doped silicon p-type or n-type substrate. An opposite doping (either n-type or p-type, respectively) is fixed to the dorsal side of the silicon substrate. This fixture process forms a p/n junction (see Fig. 2-1). In this configuration, JFETs become either p-channel or n-channel devices.

Control over the flow of electrons through the channel of the JFET is maintained with two external bias voltages. These two voltages are applied between the source and drain and between the source and gate. The first voltage directs the flow of current through the JFET's channel, while the second voltage controls the amount of current flowing through this channel.

This current flow is actually a product of another feature that is formed between the source and the drain. A reverse bias formed at the p/n junction develops a depletion zone near the junction. This depletion zone is controlled

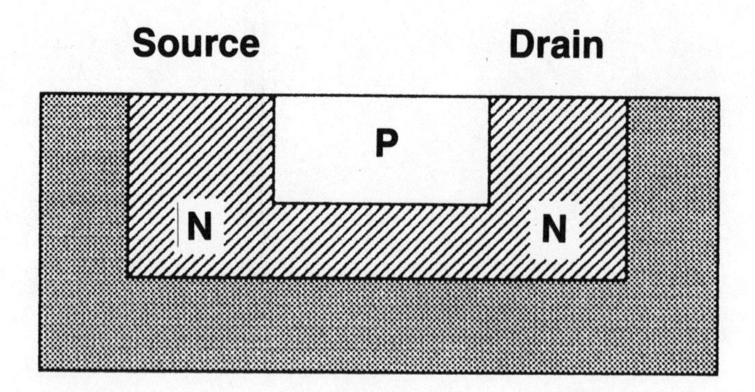

Fig. 2-1. The p/n junction of a typical GaAs FET.

by the voltage between the source and the gate. An increase in this voltage similarly increases the size of the depletion zone. Additionally, this increase in the size of the depletion zone also serves to effectively reduce the size of the channel inhibiting the flow of current. Likewise, a reduction in the size of the depletion zone increases the flow of current through the channel. Therefore, the voltage that is applied between the source and the gate is used to control the drain's current.

The second application of FETs in gallium arsenide technology is in the incorporation of multiple D-MESFETs within a DIP carrier. In this configuration, the D-MESFET can be either a single gate or a double gate with performance characteristics similar to the previously discussed silicon n-channel JFET.

Single gate D-MESFETs can be assembled inside the DIP carrier with either separate or paired sources (see Fig. 2-2). In a typical production MESFET array (e.g., GigaBit Logic's 16G020), both of these source arrangements will be found on the same IC.

Another D-MESFET configuration is the dual-gate array with two gates applied to each source/drain FET (see Fig. 2-3). One of these gates is fixed near the source, while the other is located nearer the drain. By using these dual gates, different operating voltages can be used for reducing circuit gain, producing high output impedance, and increasing reverse isolation.

A recent offshoot from GaAs FETs is the GaAs-on-silicon device. Fabricated as a complex sandwich on small wafers (less than 6 inches), GaAs-on-silicon FETs combine the important benefits from each technology into a single, fast, temperature- and radiation-resistant package. Typically, the development of this odd hybrid is from the epitaxial growth of GaAs on top of a silicon substrate. Unfortunately, vertical defect propagation and wafer warping make the successful marriage of GaAs to silicon a difficult proposition. Chemical trenching along with defect containment regions offer possible solutions for finally consummating this unlikely bond.

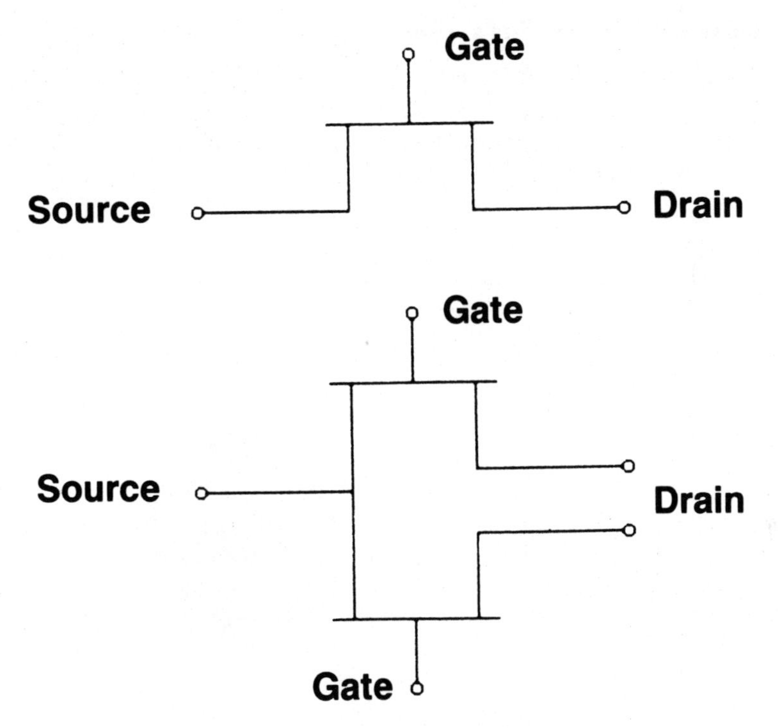

Fig. 2-2. Separate and paired sources in a D-MESFET.

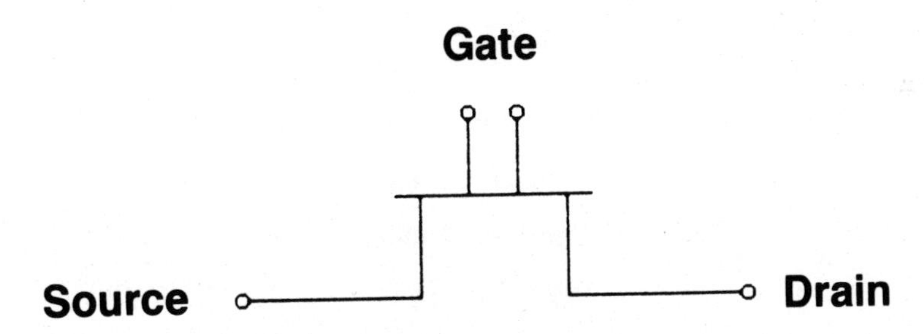

Fig. 2-3. Dual-gate array in a D-MESFET.

DUAL GATE N-CHANNEL FETs

GaAs dual-gate n-channel FETs offer low noise and high gain for rf (radio frequency) signal amplifier applications. These circuits also have a high input impedance with excellent agc (automatic gain control) capabilities.

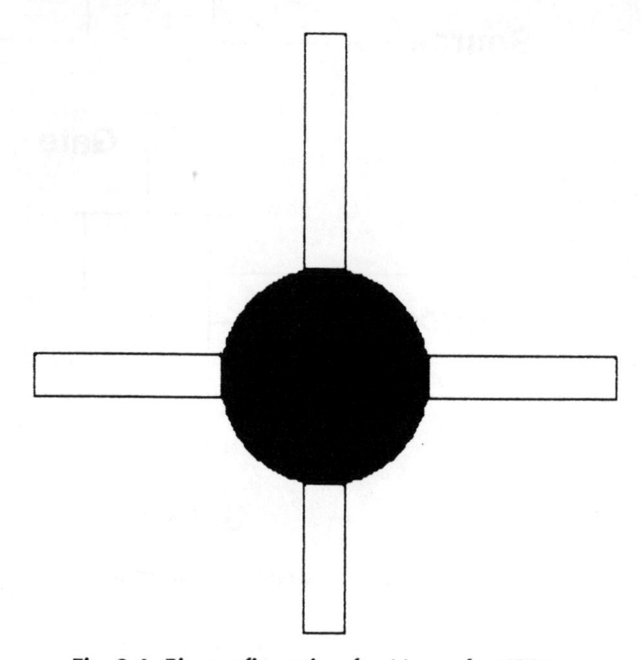

Fig. 2-4. Pin configuration for Motorola MRF966.

Device Example: Motorola MRF966

Package Configuration: Plastic SOE Case 317A

Operating Voltage: + 5.0 V

Typical Noise Figures: 1.2 dB at 1000 MHz

Gain: 15 dB at 1000 MHz

Output Level: 10 dBm

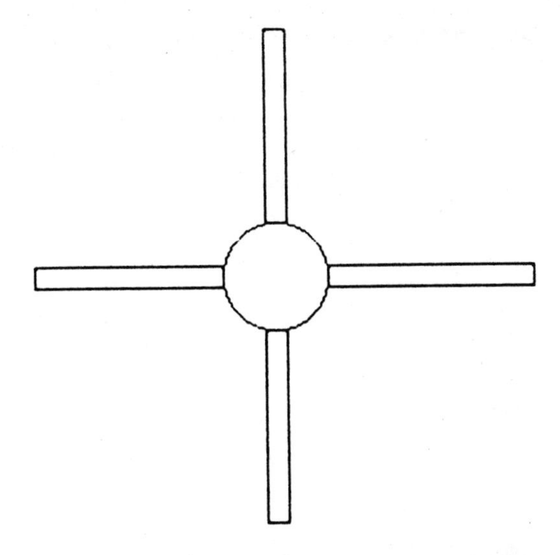

Fig. 2-5. Pin configuration for Motorola MRF967.

Device Example: Motorola MRF967

Package Configuration: Ceramic SOE Case 358

Operating Voltage: + 5.0 V

Typical Noise Figures: 1.2 dB at 1000 MHz

Gain: 13 dB at 1000 MHz

Output Level: 10 dBm

D-MESFET ARRAYS

Single- and double-gate, multiple-array D-MESFETs have a low gate capacitance paired with high output impedance. This performance combination is ideal for rf or microwave amplifiers, agc amplifiers, digital drivers, analog switches, and buffers.

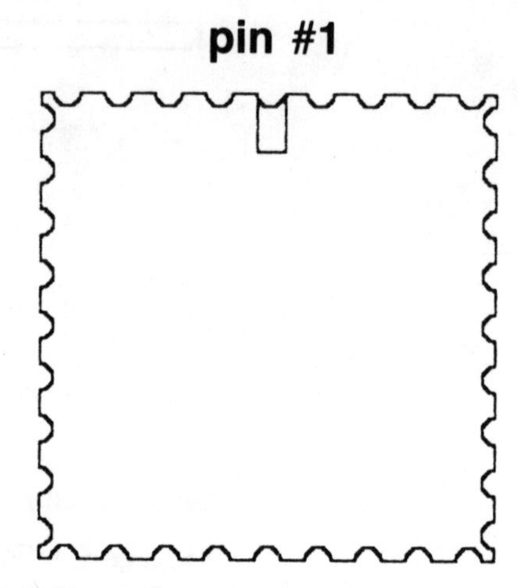

Fig. 2-6. Pin configuration for GigaBit Logic 16G020.

Device Example: GigaBit Logic 16G020

Package Configuration: 36 I/O Leadless Chip Carrier

Internal Logic Arrangement: 11 single gate D-MESFET; nine single gate and one differential source pair

Gate Voltage Operating Range: − 0.9 V to + 0.6 V

Frequency Response: 15 GHz

Pinchoff Voltage: − 0.9 V

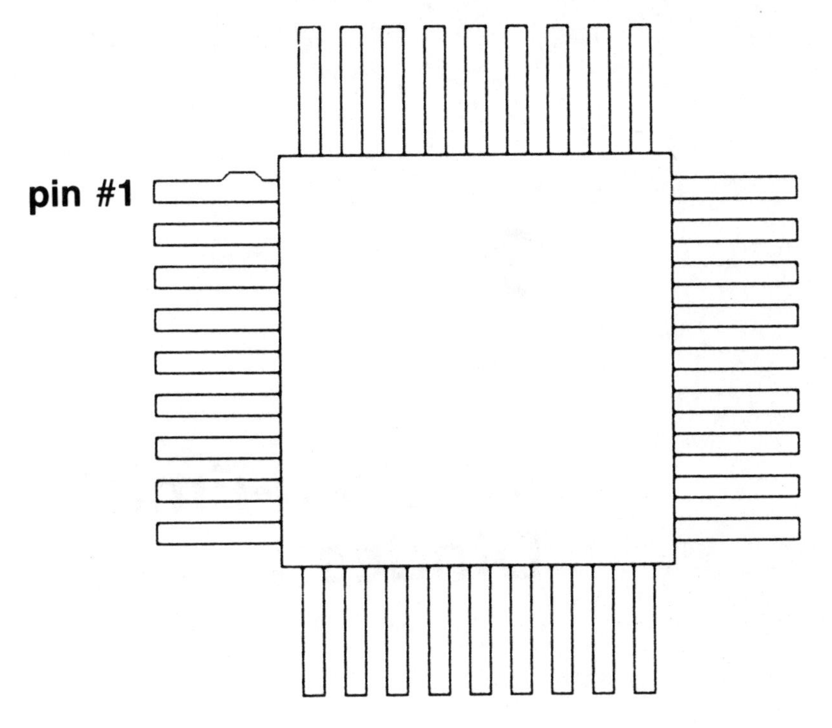

Fig. 2-7. Pin configuration for GigaBit Logic 16G021.

Device Example: GigaBit Logic 16G021

Package Configuration: 36 I/O flatpack

Internal Logic Arrangement: Eight dual-gate D-MESFETs

Gate Voltage Operating Range: Gate 1 (gV1) = – 0.9 V to + 0.6 V Gate 2 = (gV1 – 1) to (gV1 + 1)

Frequency Response: 15 GHz

Pinchoff Voltage: – 0.9 V

3

Infrared-Emitting Diodes

Like a conventional LED, an IRED is a p/n junction diode with the capacity for emitting photons. Unlike conventional LEDs, however, the photons generated by an IRED are invisible to the human eye. In either case, the production of photons is the same for both of these semiconductor p/n junction devices.

During operation, any LED (whether conventional or IR) must have electrons injected into its n-type material through a forward-bias current. As these electrons are injected into this layer, they move towards the electron-poor p-type layer of the diode. The positive charge that forms the p-type material forces holes towards the n-type layer. The result is the release of photons when the electrons and holes arrive at the p/n junction.

This electron/hole photon radiation has a wavelength from 900 to 1000 nanometers with GaAs-based LEDs. By emitting these wavelengths of light, GaAs LEDs generate light in the infrared portions of the visible light spectrum. In contrast, other visible light LEDs produce wavelengths from 550 to 900 nm. These shorter wavelengths are excited by phosphide (or aluminum) impurities that are added to the GaAs substrate.

A typical GaAs IRED has a construction architecture that is virtually identical to that of a visible light LED (see Fig. 3-1). Starting with a basal n-type GaAs substrate, an insulative coating is applied over this lower surface. A p-type impurity is injected into the epitaxial layer under this insulation. This impurity is responsible for the wavelength of photon that is emitted by the

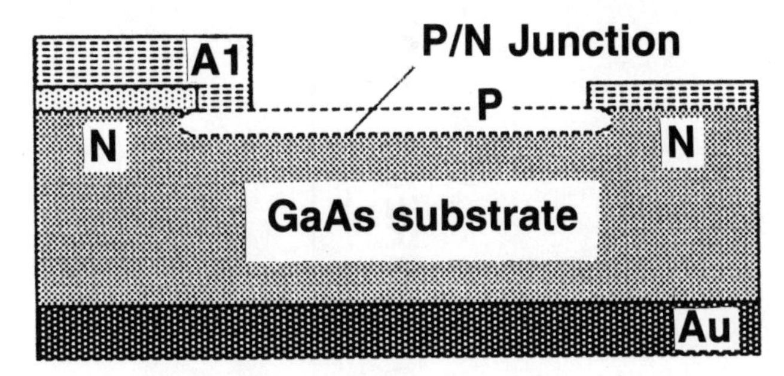

Fig. 3-1. An IRED circuit.

LED. Finally, a small opening is etched into the p/n junction for facilitating the escape of photons.

After this small GaAs wafer has been fabricated, cathode and anode connections are made to the bottom and top of the LED wafer, respectively, and the entire assembly is placed inside a plastic housing (see Fig. 3-2). Typically, two features are standard in the IRED housing. First, the cathode lead of the LED is marked by a flat "side" on the ordinarily round housing base (this indicator is usually supplemented by a shorter cathode lead length).

The second standard feature of the IRED housing is the inclusion of a plastic lens. This lens concentrates and magnifies the emitted photons into an optimal beam of infrared light. In IREDs, this lens is a clear plastic undiffused spherical dome featuring deep GaAs wafer mounting inside the housing that focuses the IR radiation into a tight concentrated beam.

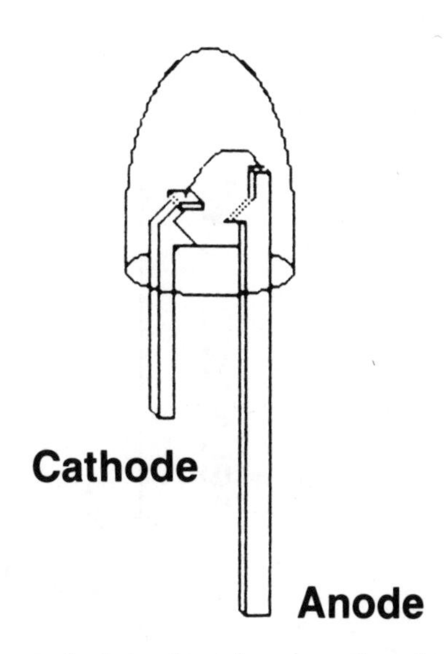

Fig. 3-2. Cathode and anode connections for an IRED.

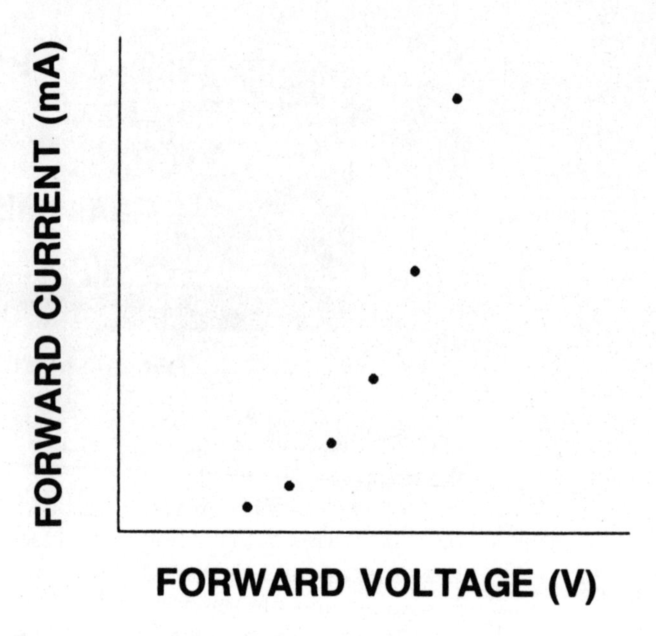

Fig. 3-3. Forward current versus forward voltage for an IRED.

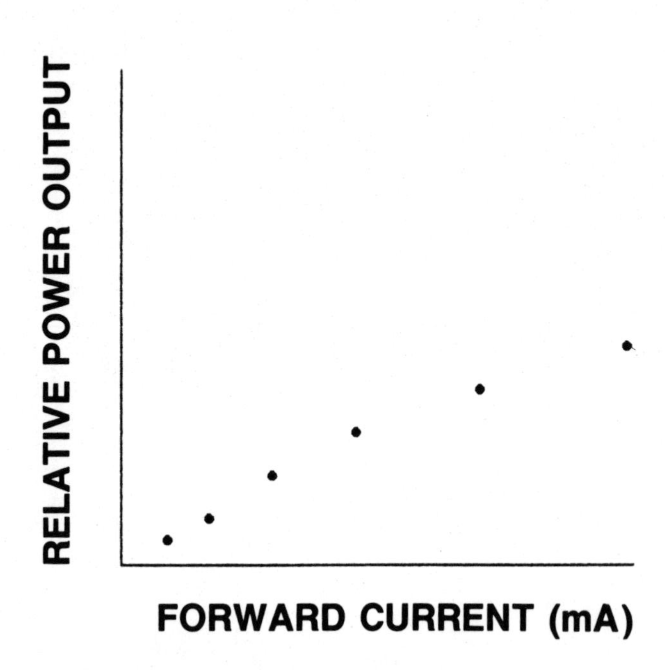

Fig. 3-4. Relative power output versus forward current for an IRED.

Photon radiation from a GaAs IRED is determined by the combination of forward current and forward voltage. In this regard, the forward bias must reach a voltage level above + 0.9 V before forward current develops a significant flow. Above this voltage level, current increases dramatically following a similar rise in voltage. This voltage level usually peaks at a fixed range between + 1.2 V to + 1.6 V (see Fig. 3-3).

Coupled to this forward current rise is the increase in the relative power output (see Fig. 3-4). The relative power output of an IRED increases linearly against the forward current. A typical relative power output for a forward current of 20 mA is 1.2 mW. Doubling the forward current to 40 mA raises the relative power output to its maximum of 2 mW.

One final factor affecting the performance of the IRED is the relative photon intensity. This level is determined by the angular displacement of the radiated light rays. The loss of these radiated light rays results in a reduction in the relative photon intensity. A standard IRED retains an 80 percent intensity level over a 10-degree-from-center angular displacement. Increasing the angular displacement to 35 degrees reduces the photon intensity to 50 percent of full intensity.

IREDS

Packaged with a wide variety of different angular displacements and operating under a broad range of power requirements, IREDs are well-suited to voice-modulated communicators, fiber-optic drivers, and open-air sensors. These emitters are designed to radiate photons in the infrared spectrum between 800 and 900 nm.

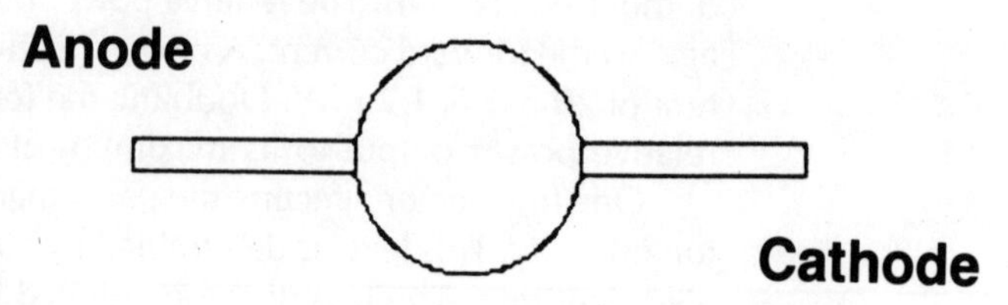

Fig. 3-5. Pin assignments for Motorola MLED15.

Product Example: Motorola MLED15

Package Configuration: Plastic case 173-01

Material Composition: GaAs

Emission Wavelength: 930 nm

Typical Radiant Power Output: 13.0 mW with a forward current of 30 mA

Angular Displacement for 50% Power Output: 145 degrees

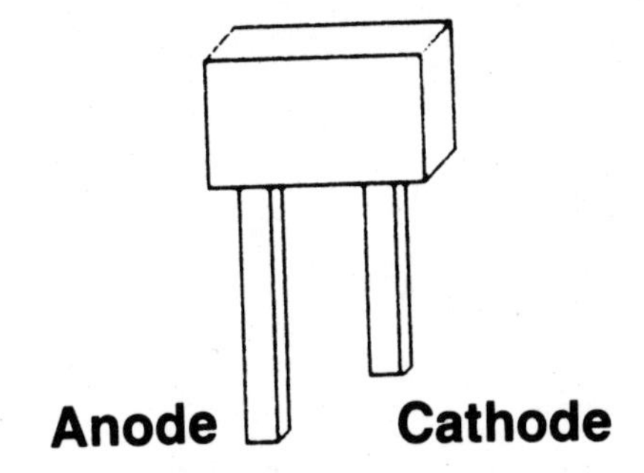

Fig. 3-6. Pin assignments for Motorola MLED71.

Product Example: Motorola MLED71

Package Configuration: Plastic case 349-01

Material Composition: GaAs

Emission Wavelength: 940 nm

Typical Radiant Power Output: 25.0 mW with a forward current of 50 mA

Angular Displacement for 50% Power Output: 60 degrees

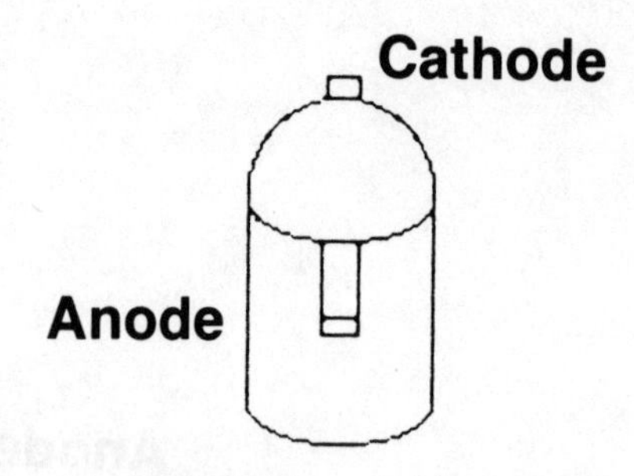

Fig. 3-7. Pin assignments for Motorola MLED910.

Product Example: Motorola MLED910

Package Configuration: Metal case 81A-06

Material Composition: GaAs

Emission Wavelength: 900 nm

Typical Radiant Power Output: 1.5 mW with a forward current of 50 mA

Angular Displacement for 50% Power Output: 30 degrees

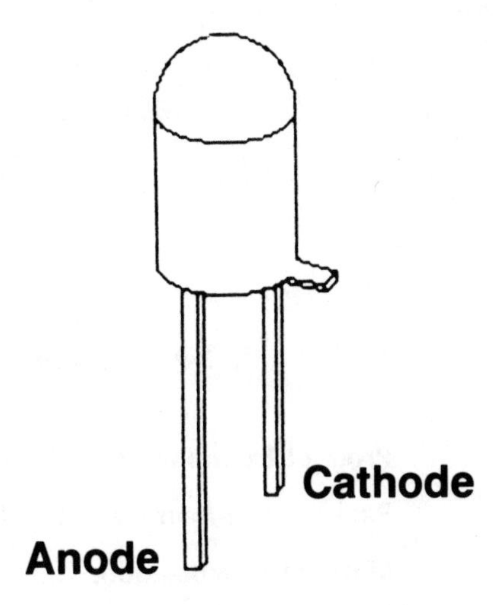

Fig. 3-8. Pin assignments for Motorola MLED930.

Product Example: Motorola MLED930

Package Configuration: Metal case 209-01

Material Composition: GaAs

Emission Wavelength: 900 nm

Typical Radiant Power Output: 6.5 mW with a forward current of 100 mA

Angular Displacement for 50% Power Output: 30 degrees

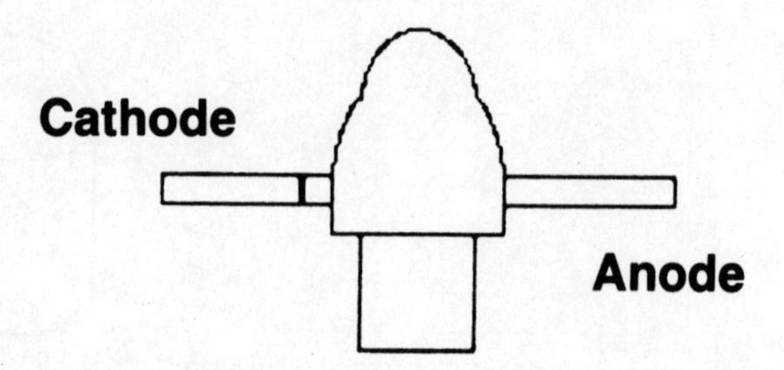

Fig. 3-9. Pin assignments for RCA SK2005 and RCA SK2006.

Product Example: RCA SK2005 and RCA SK2006

Package Configuration: L-008

Material Composition: GaAs

Emission Wavelength: 940 nm

Typical Radiant Power
Output: RCA SK2005: 1.5 mW with a forward current of 50 mA
RCA SK2005: 0.8 mW with a forward current of 50 mA

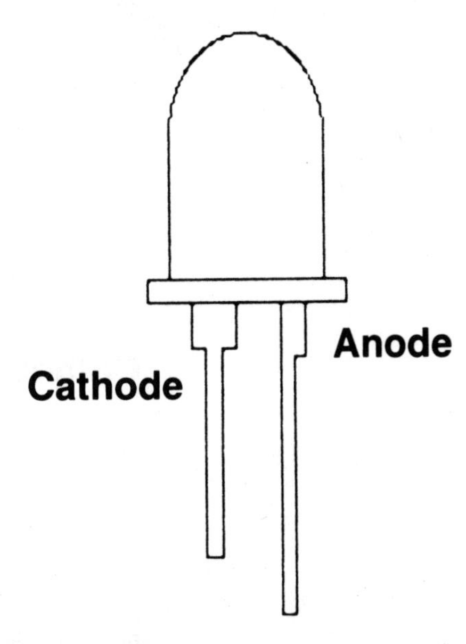

Fig. 3-10. Pin assignments for RCA SK2027.

Product Example: RCA SK2027

Package Configuration: L-009

Material Composition: GaAs

Emission Wavelength: 940 nm

Typical Radiant Power Output: 1.0 mW with a forward current of 100 mA

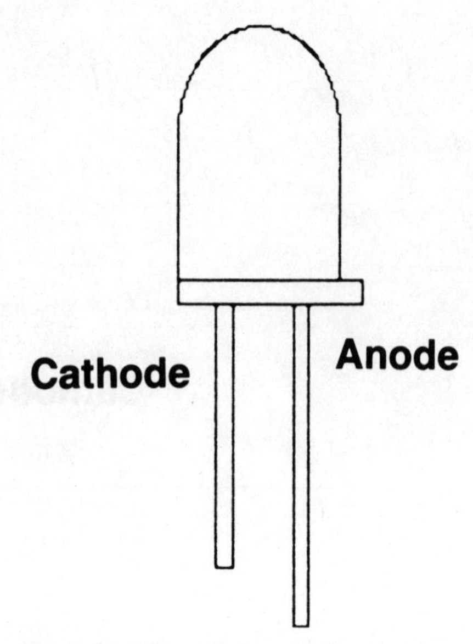

Fig. 3-11. Pin assignments for TI TIL32.

Product Example: Texas Instruments TIL32

Package Configuration: T-1

Material Composition: GaAs

Emission Wavelength: 940 nm

Typical Radiant Power Output: 1.2 mW with a forward current of 20 mA

Angular Displacement for 50% Power Output: 35 degrees

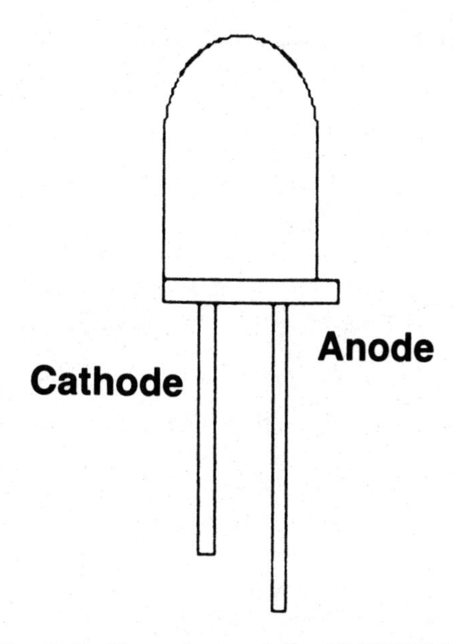

Fig. 3-12. Pin assignments for TI TIL906-1.

Product Example: Texas Instruments TIL906-1

Package Configuration: T-1

Material Composition: GaAlAs

Emission Wavelength: 880 nm

Typical Radiant Power Output: 1.5 mW with a forward current of 20 mA

Angular Displacement for 50% Power Output: 20 degrees

BAR CODE READER

One valuable application for photon emitter LEDs is in *bar code readers* or *scanners*. Generally, these emitters are paired with a spectrally-matched photodetector and housed together in a sealed package. In this configuration, the beam from the emitter is focused on a floating spot directly in front of the detector's region of greatest sensitivity.

This project, the Bar Code Reader, is based on a singular, high-resolution bar code emitter/detector pair. Unlike the other LEDs discussed in this chapter, however, the emitter in this pattern recognition system radiates visible red light.

Central to the Bar Code Reader is the Hewlett-Packard HBCS-1100 High Resolution Optical Reflector Sensor. This sensor is an integrated emitter/detector optical pattern recognition package. The emitter portion of this sensor consists of an LED radiating a 700 nm wavelength. This photon emission is passed through a bifurcated aspheric lens that focuses the light beam 4.27 mm in front of the detector. The detector is a photodiode with an internal npn high-gain transistor amplifier. This amplification section of the detector can be bypassed for direct photodiode output.

Construction Notes

The elements needed for building the Bar Code Reader are provided in Fig. 3-13. This assembly is a convenient means for interfacing the Bar Code Reader directly to a microcomputer.

Once you have built the Bar Code Reader, a test must be conducted for determining its compatibility with interpreting the bar code data formats (e.g., Universal Product Code, or UPC). The computer software listing in Fig. 3-14

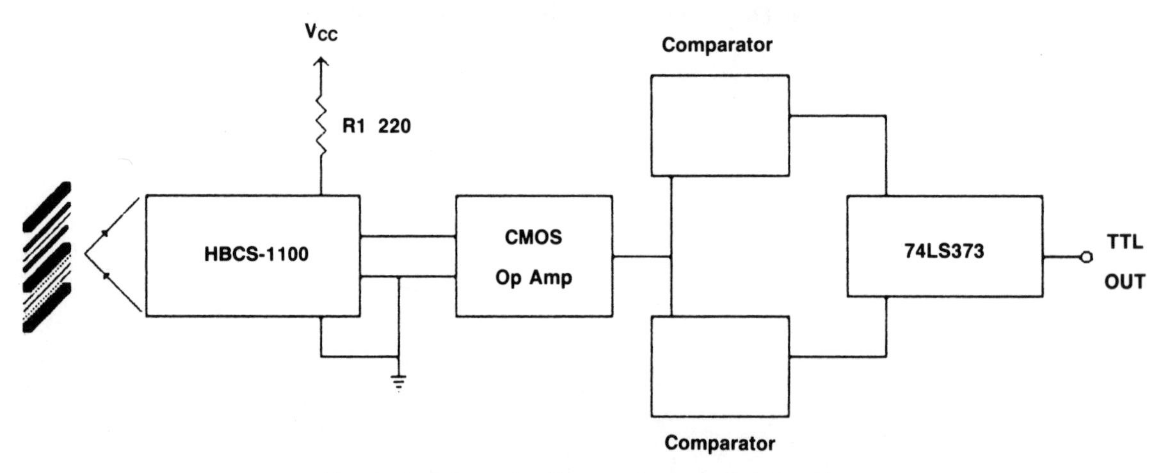

Fig. 3-13. Schematic diagram for Bar Code Reader.

is a simple bar code generation program written in 8088 machine code for the IBM PC computer. This program takes a character string input and outputs a comparable bar code on a connected Epson dot-matrix printer (e.g., FX family). An example of this program's output is shown at the end of the program.

Fig. 3-14. 8088 machine code program for generating bar codes.

```
C>
5FF0:0160 A19610        MOV     AX,[1096]
5FF0:0163 00A39601      ADD     [BP+DI+0196],AH
5FF0:0167 00B39608      ADD     [BP+DI+0896],DH
5FF0:016B 00B49608      ADD     [SI+0896],DH
5FF0:016F 00C1          ADD     CL,AL
5FF0:0171 96            XCHG    SI,AX
5FF0:0172 43            INC     BX
5FF0:0173 00C9          ADD     CL,CL
5FF0:0175 96            XCHG    SI,AX
5FF0:0176 0000          ADD     [BX+SI],AL
5FF0:0178 0C97          OR      AL,97
5FF0:017A 0000          ADD     [BX+SI],AL
5FF0:017C 0C97          OR      AL,97
5FF0:017E 0C00          OR      AL,00
5FF0:0180 0C97          OR      AL,97
5FF0:0182 16            PUSH    SS
5FF0:0183 0018          ADD     [BX+SI],BL
5FF0:0185 97            XCHG    DI,AX
5FF0:0186 0000          ADD     [BX+SI],AL
5FF0:0188 30960000      XOR     [BP+0000],DL
5FF0:018C 1001          ADC     [BX+DI],AL
5FF0:018E 0000          ADD     [BX+SI],AL
5FF0:0190 1001          ADC     [BX+DI],AL
5FF0:0192 0000          ADD     [BX+SI],AL
5FF0:0194 1001          ADC     [BX+DI],AL
5FF0:0196 0000          ADD     [BX+SI],AL
5FF0:0198 1001          ADC     [BX+DI],AL
5FF0:019A 3C00          CMP     AL,00
5FF0:019C 8C01          MOV     [BX+DI],ES
5FF0:019E 1000          ADC     [BX+SI],AL
5FF0:01A0 C8            DB      C8
5FF0:01A1 0102          ADD     [BP+SI],AX
5FF0:01A3 0020          ADD     [BX+SI],AH
5FF0:01A5 0100          ADD     [BX+SI],AX
5FF0:01A7 008C012E      ADD     [SI+2E01],CL
5FF0:01AB 8C1EB200      MOV     [00B2],DS
```

```
5FF0:01AF 2E            CS:
5FF0:01B0 C706AE000000  MOV     WORD PTR [00AE],0000
5FF0:01B6 E85405        CALL    070D
5FF0:01B9 E88800        CALL    0244
5FF0:01BC E81717        CALL    18D6
5FF0:01BF E8AE17        CALL    1970
5FF0:01C2 E81100        CALL    01D6
5FF0:01C5 2E            CS:
5FF0:01C6 8E1E6401      MOV     DS,[0164]
5FF0:01CA FF360200      PUSH    [0002]
5FF0:01CE E89F17        CALL    1970
5FF0:01D1 58            POP     AX
5FF0:01D2 B44C          MOV     AH,4C
5FF0:01D4 CD21          INT     21
5FF0:01D6 E8A700        CALL    0280
5FF0:01D9 E8D600        CALL    02B2
5FF0:01DC 7253          JB      0231
5FF0:01DE E8C401        CALL    03A5
5FF0:01E1 E84802        CALL    042C
5FF0:01E4 E86C02        CALL    0453
5FF0:01E7 E8C61F        CALL    21B0
5FF0:01EA E8252F        CALL    3112
5FF0:01ED E85201        CALL    0342
5FF0:01F0 0E            PUSH    CS
5FF0:01F1 1F            POP     DS
5FF0:01F2 A15601        MOV     AX,[0156]
5FF0:01F5 B104          MOV     CL,04
5FF0:01F7 D3E0          SHL     AX,CL
5FF0:01F9 A3CE00        MOV     [00CE],AX
5FF0:01FC B88000        MOV     AX,0080
5FF0:01FF A3D200        MOV     [00D2],AX
5FF0:0202 8E1E6401      MOV     DS,[0164]
5FF0:0206 E84804        CALL    0651
5FF0:0209 E87A01        CALL    0386
5FF0:020C 0E            PUSH    CS
5FF0:020D 1F            POP     DS
5FF0:020E 8926C200      MOV     [00C2],SP
5FF0:0212 8C16C400      MOV     [00C4],SS
5FF0:0216 FA            CLI
5FF0:0217 8E165801      MOV     SS,[0158]
5FF0:021B 8B26CE00      MOV     SP,[00CE]
5FF0:021F FC            CLD
5FF0:0220 FB            STI
5FF0:0221 33C9          XOR     CX,CX
```

```
5FF0:0223 33FF          XOR     DI,DI
5FF0:0225 33ED          XOR     BP,BP
5FF0:0227 2E            CS:
5FF0:0228 8E1E6401      MOV     DS,[0164]
5FF0:022C 2E            CS:
5FF0:022D FF2ED400      JMP     FAR [00D4]
5FF0:0231 E89547        CALL    49C9
5FF0:0234 2E            CS:
5FF0:0235 833EAE0000    CMP     WORD PTR [00AE],+00
5FF0:023A 7401          JZ      023D
5FF0:023C C3            RET
5FF0:023D E83017        CALL    1970
5FF0:0240 32E4          XOR     AH,AH
5FF0:0242 CD21          INT     21
5FF0:0244 8CC8          MOV     AX,CS
5FF0:0246 2E            CS:
5FF0:0247 0306BA00      ADD     AX,[00BA]
5FF0:024B 2D8200        SUB     AX,0082
5FF0:024E 8ED8          MOV     DS,AX
5FF0:0250 A17C00        MOV     AX,[007C]
5FF0:0253 8BD0          MOV     DX,AX
5FF0:0255 D1E0          SHL     AX,1
5FF0:0257 03C2          ADD     AX,DX
5FF0:0259 E83018        CALL    1A8C
5FF0:025C 03067E00      ADD     AX,[007E]
5FF0:0260 03063400      ADD     AX,[0034]
5FF0:0264 03063800      ADD     AX,[0038]
5FF0:0268 8BD0          MOV     DX,AX
5FF0:026A B8A052        MOV     AX,52A0
5FF0:026D E81C18        CALL    1A8C
5FF0:0270 8CCB          MOV     BX,CS
5FF0:0272 03C3          ADD     AX,BX
5FF0:0274 8ED8          MOV     DS,AX
5FF0:0276 058200        ADD     AX,0082
5FF0:0279 8EC0          MOV     ES,AX
5FF0:027B 8BC2          MOV     AX,DX
5FF0:027D E96617        JMP     19E6
5FF0:0280 8CC8          MOV     AX,CS
5FF0:0282 8BD0          MOV     DX,AX
5FF0:0284 2E            CS:
5FF0:0285 0316BA00      ADD     DX,[00BA]
5FF0:0289 33F6          XOR     SI,SI
5FF0:028B 8EDA          MOV     DS,DX
5FF0:028D BF0A01        MOV     DI,010A
```

```
5FF0:0290 8EC0         MOV    ES,AX
5FF0:0292 B94000       MOV    CX,0040
5FF0:0295 FC           CLD
5FF0:0296 F3           REPZ
5FF0:0297 A5           MOVSW
5FF0:0298 8ED8         MOV    DS,AX
5FF0:029A 01168801     ADD    [0188],DX
5FF0:029E 83C208       ADD    DX,+08
5FF0:02A1 03164601     ADD    DX,[0146]
5FF0:02A5 03165801     ADD    DX,[0158]
5FF0:02A9 03165601     ADD    DX,[0156]
5FF0:02AD 01168401     ADD    [0184],DX
5FF0:02B1 C3           RET
5FF0:02B2 E87E00       CALL   0333
5FF0:02B5 7208         JB     02BF
5FF0:02B7 E84300       CALL   02FD
5FF0:02BA 7203         JB     02BF
5FF0:02BC E80100       CALL   02C0
5FF0:02BF C3           RET
5FF0:02C0 2E           CS:
5FF0:02C1 F706A4000100 TEST   WORD PTR [00A4],0001
5FF0:02C7 750D         JNZ    02D6
5FF0:02C9 2E           CS:
5FF0:02CA F70616010080 TEST   WORD PTR [0116],8000
5FF0:02D0 7404         JZ     02D6
5FF0:02D2 B80101       MOV    AX,0101
5FF0:02D5 F9           STC
5FF0:02D6 C3           RET
5FF0:02D7 2E           CS:
5FF0:02D8 0B06FC00     OR     AX,[00FC]
5FF0:02DC 2E           CS:
5FF0:02DD A3F800       MOV    [00F8],AX
5FF0:02E0 2E           CS:
5FF0:02E1 F706A4000100 TEST   WORD PTR [00A4],0001
5FF0:02E7 740A         JZ     02F3
5FF0:02E9 DBE3                FINIT
5FF0:02EB 9B           WAIT
5FF0:02EC 2E           CS:
5FF0:02ED D92EF800            FLDCW   [00F8]
5FF0:02F1 9B           WAIT
5FF0:02F2 C3           RET
5FF0:02F3 CD37         INT    37
5FF0:02F5 E3CD         JCXZ   02C4
5FF0:02F7 3C99         CMP    AL,99
```

```
5FF0:02F9 2E            CS:
5FF0:02FA F8            CLC
5FF0:02FB 00C3          ADD     BL,AL
5FF0:02FD B44A          MOV     AH,4A
5FF0:02FF BBFFFF        MOV     BX,FFFF
5FF0:0302 2E            CS:
5FF0:0303 8E06B200      MOV     ES,[00B2]
5FF0:0307 CD21          INT     21
5FF0:0309 B44A          MOV     AH,4A
5FF0:030B CD21          INT     21
5FF0:030D 2E            CS:
5FF0:030E 031EB200      ADD     BX,[00B2]
5FF0:0312 2E            CS:
5FF0:0313 3B1E1C01      CMP     BX,[011C]
5FF0:0317 7605          JBE     031E
5FF0:0319 2E            CS:
5FF0:031A 8B1E1C01      MOV     BX,[011C]
5FF0:031E 2E            CS:
5FF0:031F 891E0001      MOV     [0100],BX
5FF0:0323 2E            CS:
5FF0:0324 891E0201      MOV     [0102],BX
5FF0:0328 2E            CS:
5FF0:0329 3B1E8401      CMP     BX,[0184]
5FF0:032D 7303          JNB     0332
5FF0:032F B80201        MOV     AX,0102
5FF0:0332 C3            RET
5FF0:0333 2E            CS:
5FF0:0334 A1AC00        MOV     AX,[00AC]
5FF0:0337 3C02          CMP     AL,02
5FF0:0339 7202          JB      033D
5FF0:033B F8            CLC
5FF0:033C C3            RET
5FF0:033D B80001        MOV     AX,0100
5FF0:0340 F9            STC
5FF0:0341 C3            RET
5FF0:0342 2E            CS:
5FF0:0343 C70696000000  MOV     WORD PTR [0096],0000
5FF0:0349 CDEC          INT     EC
5FF0:034B 6A            DB      6A
5FF0:034C CDED          INT     ED
5FF0:034E 2ACD          SUB     CL,CH
5FF0:0350 3C9F          CMP     AL,9F
5FF0:0352 1E            PUSH    DS
5FF0:0353 96            XCHG    SI,AX
```

```
5FF0:0354 00C3          ADD    BL,AL
5FF0:0356 2E            CS:
5FF0:0357 8B168001      MOV    DX,[0180]
5FF0:035B 2E            CS:
5FF0:035C 8B0E0201      MOV    CX,[0102]
5FF0:0360 8BC2          MOV    AX,DX
5FF0:0362 050010        ADD    AX,1000
5FF0:0365 3BC1          CMP    AX,CX
5FF0:0367 7602          JBE    036B
5FF0:0369 8BC1          MOV    AX,CX
5FF0:036B 2E            CS:
5FF0:036C A30401        MOV    [0104],AX
5FF0:036F 2BC2          SUB    AX,DX
5FF0:0371 B104          MOV    CL,04
5FF0:0373 D3E0          SHL    AX,CL
5FF0:0375 8EDA          MOV    DS,DX
5FF0:0377 2B060200      SUB    AX,[0002]
5FF0:037B A30000        MOV    [0000],AX
5FF0:037E 2E            CS:
5FF0:037F A16401        MOV    AX,[0164]
5FF0:0382 A30400        MOV    [0004],AX
5FF0:0385 C3            RET
5FF0:0386 E8CDFF        CALL   0356
5FF0:0389 B80800        MOV    AX,0008
5FF0:038C B90080        MOV    CX,8000
5FF0:038F 890E0800      MOV    [0008],CX
5FF0:0393 A30A00        MOV    [000A],AX
5FF0:0396 A30C00        MOV    [000C],AX
5FF0:0399 890E0E00      MOV    [000E],CX
5FF0:039D 2E            CS:
5FF0:039E C70606010000  MOV    WORD PTR [0106],0000
5FF0:03A4 C3            RET
5FF0:03A5 8CC8          MOV    AX,CS
5FF0:03A7 2E            CS:
5FF0:03A8 01064001      ADD    [0140],AX
5FF0:03AC 2E            CS:
5FF0:03AD 01064401      ADD    [0144],AX
5FF0:03B1 2E            CS:
5FF0:03B2 0306BA00      ADD    AX,[00BA]
5FF0:03B6 050800        ADD    AX,0008
5FF0:03B9 2E            CS:
5FF0:03BA A3D600        MOV    [00D6],AX
5FF0:03BD 2E            CS:
5FF0:03BE A34801        MOV    [0148],AX
```

```
5FF0:03C1 2E            CS:
5FF0:03C2 A31E01        MOV     [011E],AX
5FF0:03C5 B90F00        MOV     CX,000F
5FF0:03C8 BF2001        MOV     DI,0120
5FF0:03CB 833D00        CMP     WORD PTR [DI],+00
5FF0:03CE 7407          JZ      03D7
5FF0:03D0 0105          ADD     [DI],AX
5FF0:03D2 83C702        ADD     DI,+02
5FF0:03D5 E2F4          LOOP    03CB
5FF0:03D7 2E            CS:
5FF0:03D8 03064601      ADD     AX,[0146]
5FF0:03DC 2E            CS:
5FF0:03DD 01064C01      ADD     [014C],AX
5FF0:03E1 2E            CS:
5FF0:03E2 01065001      ADD     [0150],AX
5FF0:03E6 2E            CS:
5FF0:03E7 01065401      ADD     [0154],AX
5FF0:03EB 2E            CS:
5FF0:03EC 01065801      ADD     [0158],AX
5FF0:03F0 2E            CS:
5FF0:03F1 A15801        MOV     AX,[0158]
5FF0:03F4 2E            CS:
5FF0:03F5 03065601      ADD     AX,[0156]
5FF0:03F9 2E            CS:
5FF0:03FA 01065C01      ADD     [015C],AX
5FF0:03FE 2E            CS:
5FF0:03FF 01066001      ADD     [0160],AX
5FF0:0403 2E            CS:
5FF0:0404 01066401      ADD     [0164],AX
5FF0:0408 2E            CS:
5FF0:0409 01066801      ADD     [0168],AX
5FF0:040D 2E            CS:
5FF0:040E 01066C01      ADD     [016C],AX
5FF0:0412 2E            CS:
5FF0:0413 01067001      ADD     [0170],AX
5FF0:0417 2E            CS:
5FF0:0418 01067401      ADD     [0174],AX
5FF0:041C 2E            CS:
5FF0:041D 01067801      ADD     [0178],AX
5FF0:0421 2E            CS:
5FF0:0422 01067C01      ADD     [017C],AX
5FF0:0426 2E            CS:
5FF0:0427 01068001      ADD     [0180],AX
5FF0:042B C3            RET
```

```
5FF0:042C 2E           CS:
5FF0:042D A14801       MOV    AX,[0148]
5FF0:0430 32FF         XOR    BH,BH
5FF0:0432 2E           CS:
5FF0:0433 8B0E8601     MOV    CX,[0186]
5FF0:0437 2E           CS:
5FF0:0438 8E1E8801     MOV    DS,[0188]
5FF0:043C 33F6         XOR    SI,SI
5FF0:043E E312         JCXZ   0452
5FF0:0440 8B3C         MOV    DI,[SI]
5FF0:0442 8A5C02       MOV    BL,[SI+02]
5FF0:0445 2E           CS:
5FF0:0446 8E871E01     MOV    ES,[BX+011E]
5FF0:044A 26           ES:
5FF0:044B 0105         ADD    [DI],AX
5FF0:044D 83C603       ADD    SI,+03
5FF0:0450 E2EE         LOOP   0440
5FF0:0452 C3           RET
5FF0:0453 2E           CS:
5FF0:0454 8B168801     MOV    DX,[0188]
5FF0:0458 BE8001       MOV    SI,0180
5FF0:045B E85900       CALL   04B7
5FF0:045E BE7001       MOV    SI,0170
5FF0:0461 E85300       CALL   04B7
5FF0:0464 BE6C01       MOV    SI,016C
5FF0:0467 E84D00       CALL   04B7
5FF0:046A BE6801       MOV    SI,0168
5FF0:046D E84700       CALL   04B7
5FF0:0470 BE6001       MOV    SI,0160
5FF0:0473 E84100       CALL   04B7
5FF0:0476 BE5C01       MOV    SI,015C
5FF0:0479 E83B00       CALL   04B7
5FF0:047C 8CC8         MOV    AX,CS
5FF0:047E 8EC0         MOV    ES,AX
5FF0:0480 33F6         XOR    SI,SI
5FF0:0482 2E           CS:
5FF0:0483 8E1E6801     MOV    DS,[0168]
5FF0:0487 BF8A01       MOV    DI,018A
5FF0:048A B90800       MOV    CX,0008
5FF0:048D FC           CLD
5FF0:048E F3           REPZ
5FF0:048F A5           MOVSW
5FF0:0490 33F6         XOR    SI,SI
5FF0:0492 2E           CS:
```

```
5FF0:0493 8E1E6C01      MOV   DS,[016C]
5FF0:0497 B90800        MOV   CX,0008
5FF0:049A F3            REPZ
5FF0:049B A5            MOVSW
5FF0:049C 8ED8          MOV   DS,AX
5FF0:049E 8B0EA201      MOV   CX,[01A2]
5FF0:04A2 8B36A401      MOV   SI,[01A4]
5FF0:04A6 8B166401      MOV   DX,[0164]
5FF0:04AA 8EC2          MOV   ES,DX
5FF0:04AC E308          JCXZ  04B6
5FF0:04AE 26            ES:
5FF0:04AF 0114          ADD   [SI],DX
5FF0:04B1 83C636        ADD   SI,+36
5FF0:04B4 E2F8          LOOP  04AE
5FF0:04B6 C3            RET
5FF0:04B7 2E            CS:
5FF0:04B8 8B44FE        MOV   AX,[SI-02]
5FF0:04BB 2BD0          SUB   DX,AX
5FF0:04BD 2E            CS:
5FF0:04BE 8E04          MOV   ES,[SI]
5FF0:04C0 8EDA          MOV   DS,DX
5FF0:04C2 E92115        JMP   19E6
5FF0:04C5 8CCB          MOV   BX,CS
5FF0:04C7 8EDB          MOV   DS,BX
5FF0:04C9 FC            CLD
5FF0:04CA 33C0          XOR   AX,AX
5FF0:04CC 8E066401      MOV   ES,[0164]
5FF0:04D0 F706FE00FFFF  TEST  WORD PTR [00FE],FFFF
5FF0:04D6 7518          JNZ   04F0
5FF0:04D8 8B0E8A01      MOV   CX,[018A]
5FF0:04DC 8B3E8C01      MOV   DI,[018C]
5FF0:04E0 D1E9          SHR   CX,1
5FF0:04E2 F3            REPZ
5FF0:04E3 AB            STOSW
5FF0:04E4 8B0E8E01      MOV   CX,[018E]
5FF0:04E8 8B3E9001      MOV   DI,[0190]
5FF0:04EC D1E9          SHR   CX,1
5FF0:04EE F3            REPZ
5FF0:04EF AB            STOSW
5FF0:04F0 8B0E9A01      MOV   CX,[019A]
5FF0:04F4 8B3E9C01      MOV   DI,[019C]
5FF0:04F8 D1E9          SHR   CX,1
5FF0:04FA F3            REPZ
5FF0:04FB AB            STOSW
```

```
5FF0:04FC 8B0E9E01      MOV   CX,[019E]
5FF0:0500 8B3EA001      MOV   DI,[01A0]
5FF0:0504 D1E9          SHR   CX,1
5FF0:0506 F3            REPZ
5FF0:0507 AB            STOSW
5FF0:0508 F706FE00FFFF  TEST  WORD PTR [00FE],FFFF
5FF0:050E 7512          JNZ   0522
5FF0:0510 8B0E9601      MOV   CX,[0196]
5FF0:0514 8B3E9801      MOV   DI,[0198]
5FF0:0518 E308          JCXZ  0522
5FF0:051A 26            ES:
5FF0:051B 8905          MOV   [DI],AX
5FF0:051D 83C736        ADD   DI,+36
5FF0:0520 E2F8          LOOP  051A
5FF0:0522 8B0EA601      MOV   CX,[01A6]
5FF0:0526 8B3EA801      MOV   DI,[01A8]
5FF0:052A E308          JCXZ  0534
5FF0:052C 26            ES:
5FF0:052D 8905          MOV   [DI],AX
5FF0:052F 83C736        ADD   DI,+36
5FF0:0532 E2F8          LOOP  052C
5FF0:0534 C4167201      LES   DX,[0172]
5FF0:0538 E82815        CALL  1A63
5FF0:053B C4167601      LES   DX,[0176]
5FF0:053F E82115        CALL  1A63
5FF0:0542 F706FE00FFFF  TEST  WORD PTR [00FE],FFFF
5FF0:0548 751C          JNZ   0566
5FF0:054A 8E067C01      MOV   ES,[017C]
5FF0:054E C5067E01      LDS   AX,[017E]
5FF0:0552 E89114        CALL  19E6
5FF0:0555 8EDB          MOV   DS,BX
5FF0:0557 2E            CS:
5FF0:0558 A17C01        MOV   AX,[017C]
5FF0:055B 2E            CS:
5FF0:055C A38001        MOV   [0180],AX
5FF0:055F 2E            CS:
5FF0:0560 C70608010000  MOV   WORD PTR [0108],0000
5FF0:0566 C3            RET
5FF0:0567 0E            PUSH  CS
5FF0:0568 1F            POP   DS
5FF0:0569 33C0          XOR   AX,AX
5FF0:056B 8706FE00      XCHG  AX,[00FE]
5FF0:056F 0BC0          OR    AX,AX
5FF0:0571 755C          JNZ   05CF
```

```
5FF0:0573 33C0          XOR    AX,AX
5FF0:0575 87060601      XCHG   AX,[0106]
5FF0:0579 8B0E8001      MOV    CX,[0180]
5FF0:057D 8ED9          MOV    DS,CX
5FF0:057F 8EC1          MOV    ES,CX
5FF0:0581 0BC0          OR     AX,AX
5FF0:0583 7415          JZ     059A
5FF0:0585 8BF0          MOV    SI,AX
5FF0:0587 F7D8          NEG    AX
5FF0:0589 E8EF15        CALL   1B7B
5FF0:058C BF1000        MOV    DI,0010
5FF0:058F 03F7          ADD    SI,DI
5FF0:0591 8B0C          MOV    CX,[SI]
5FF0:0593 81E1FF7F      AND    CX,7FFF
5FF0:0597 FC            CLD
5FF0:0598 F3            REPZ
5FF0:0599 A4            MOVSB
5FF0:059A 2E            CS:
5FF0:059B 8B168001      MOV    DX,[0180]
5FF0:059F 2E            CS:
5FF0:05A0 A10401        MOV    AX,[0104]
5FF0:05A3 2BC2          SUB    AX,DX
5FF0:05A5 B104          MOV    CL,04
5FF0:05A7 D3E0          SHL    AX,CL
5FF0:05A9 8B0E1000      MOV    CX,[0010]
5FF0:05AD 81E1FF7F      AND    CX,7FFF
5FF0:05B1 83C118        ADD    CX,+18
5FF0:05B4 2BC1          SUB    AX,CX
5FF0:05B6 A30000        MOV    [0000],AX
5FF0:05B9 890E0200      MOV    [0002],CX
5FF0:05BD C70606000000  MOV    WORD PTR [0006],0000
5FF0:05C3 C7060A000800  MOV    WORD PTR [000A],0008
5FF0:05C9 C7060C000800  MOV    WORD PTR [000C],0008
5FF0:05CF C3            RET
5FF0:05D0 0E            PUSH   CS
5FF0:05D1 1F            POP    DS
5FF0:05D2 F606DC3080    TEST   BYTE PTR [30DC],80
5FF0:05D7 7503          JNZ    05DC
5FF0:05D9 E88C2C        CALL   3268
5FF0:05DC C70632540000  MOV    WORD PTR [5432],0000
5FF0:05E2 C70634540400  MOV    WORD PTR [5434],0004
5FF0:05E8 C70636540400  MOV    WORD PTR [5436],0004
5FF0:05EE C70638540000  MOV    WORD PTR [5438],0000
5FF0:05F4 C7063A547800  MOV    WORD PTR [543A],0078
```

```
5FF0:05FA C7063C548D04  MOV     WORD PTR [543C],048D
5FF0:0600 C7063E540300  MOV     WORD PTR [543E],0003
5FF0:0606 C3            RET
5FF0:0607 8CC8          MOV     AX,CS
5FF0:0609 8ED8          MOV     DS,AX
5FF0:060B 8EC0          MOV     ES,AX
5FF0:060D 33C0          XOR     AX,AX
5FF0:060F A3D800        MOV     [00D8],AX
5FF0:0612 A3DA00        MOV     [00DA],AX
5FF0:0615 A3EA00        MOV     [00EA],AX
5FF0:0618 C706EE00FFFF  MOV     WORD PTR [00EE],FFFF
5FF0:061E A31A54        MOV     [541A],AX
5FF0:0621 FC            CLD
5FF0:0622 B9F300        MOV     CX,00F3
5FF0:0625 BFA052        MOV     DI,52A0
5FF0:0628 F3            REPZ
5FF0:0629 AB            STOSW
5FF0:062A 48            DEC     AX
5FF0:062B B95600        MOV     CX,0056
5FF0:062E BF8654        MOV     DI,5486
5FF0:0631 F3            REPZ
5FF0:0632 AB            STOSW
5FF0:0633 33C0          XOR     AX,AX
5FF0:0635 BB5000        MOV     BX,0050
5FF0:0638 B96400        MOV     CX,0064
5FF0:063B B620          MOV     DH,20
5FF0:063D BF9857        MOV     DI,5798
5FF0:0640 AB            STOSW
5FF0:0641 32E6          XOR     AH,DH
5FF0:0643 AB            STOSW
5FF0:0644 32E6          XOR     AH,DH
5FF0:0646 03C3          ADD     AX,BX
5FF0:0648 E2F6          LOOP    0640
5FF0:064A C70620540400  MOV     WORD PTR [5420],0004
5FF0:0650 C3            RET
5FF0:0651 FC            CLD
5FF0:0652 53            PUSH    BX
5FF0:0653 1E            PUSH    DS
5FF0:0654 07            POP     ES
5FF0:0655 B020          MOV     AL,20
5FF0:0657 B91000        MOV     CX,0010
5FF0:065A BF0800        MOV     DI,0008
5FF0:065D F3            REPZ
5FF0:065E AA            STOSB
```

```
5FF0:065F 33C0          XOR     AX,AX
5FF0:0661 B90A00        MOV     CX,000A
5FF0:0664 BFD000        MOV     DI,00D0
5FF0:0667 F3            REPZ
5FF0:0668 AB            STOSW
5FF0:0669 2E            CS:
5FF0:066A 8B1ED200      MOV     BX,[00D2]
5FF0:066E 2E            CS:
5FF0:066F 8B0E8001      MOV     CX,[0180]
5FF0:0673 890E0000      MOV     [0000],CX
5FF0:0677 A30200        MOV     [0002],AX
5FF0:067A A30400        MOV     [0004],AX
5FF0:067D A31800        MOV     [0018],AX
5FF0:0680 8C1E1C00      MOV     [001C],DS
5FF0:0684 2E            CS:
5FF0:0685 A1FC00        MOV     AX,[00FC]
5FF0:0688 A31E00        MOV     [001E],AX
5FF0:068B A32000        MOV     [0020],AX
5FF0:068E A32200        MOV     [0022],AX
5FF0:0691 A32400        MOV     [0024],AX
5FF0:0694 A32600        MOV     [0026],AX
5FF0:0697 33C0          XOR     AX,AX
5FF0:0699 810E1E003F03  OR      WORD PTR [001E],033F
5FF0:069F 810E20003203  OR      WORD PTR [0020],0332
5FF0:06A5 810E2200320B  OR      WORD PTR [0022],0B32
5FF0:06AB 810E2400320F  OR      WORD PTR [0024],0F32
5FF0:06B1 810E26003207  OR      WORD PTR [0026],0732
5FF0:06B7 A35C00        MOV     [005C],AX
5FF0:06BA A35E00        MOV     [005E],AX
5FF0:06BD A36400        MOV     [0064],AX
5FF0:06C0 A36600        MOV     [0066],AX
5FF0:06C3 A36800        MOV     [0068],AX
5FF0:06C6 A36E00        MOV     [006E],AX
5FF0:06C9 A37000        MOV     [0070],AX
5FF0:06CC A37200        MOV     [0072],AX
5FF0:06CF A37400        MOV     [0074],AX
5FF0:06D2 A37600        MOV     [0076],AX
5FF0:06D5 A37800        MOV     [0078],AX
5FF0:06D8 891E8600      MOV     [0086],BX
5FF0:06DC 2E            CS:
5FF0:06DD A10001        MOV     AX,[0100]
5FF0:06E0 2E            CS:
5FF0:06E1 F7268A18      MUL     WORD PTR [188A]
5FF0:06E5 2D0100        SUB     AX,0001
```

```
5FF0:06E8 7301        JNB    06EB
5FF0:06EA 4A          DEC    DX
5FF0:06EB A36A00      MOV    [006A],AX
5FF0:06EE 89166C00    MOV    [006C],DX
5FF0:06F2 5B          POP    BX
5FF0:06F3 C3          RET
5FF0:06F4 58          POP    AX
5FF0:06F5 FA          CLI
5FF0:06F6 2E          CS:
5FF0:06F7 89268000    MOV    [0080],SP
5FF0:06FB 2E          CS:
5FF0:06FC 8C168200    MOV    [0082],SS
5FF0:0700 2E          CS:
5FF0:0701 8E16C400    MOV    SS,[00C4]
5FF0:0705 2E          CS:
5FF0:0706 8B26C200    MOV    SP,[00C2]
5FF0:070A FB          STI
5FF0:070B FFE0        JMP    AX
5FF0:070D FC          CLD
5FF0:070E 33C0        XOR    AX,AX
5FF0:0710 B90700      MOV    CX,0007
5FF0:0713 BFA000      MOV    DI,00A0
5FF0:0716 0E          PUSH   CS
5FF0:0717 07          POP    ES
5FF0:0718 F3          REPZ
5FF0:0719 AB          STOSW
5FF0:071A E81600      CALL   0733
5FF0:071D B430        MOV    AH,30
5FF0:071F CD21        INT    21
5FF0:0721 2E          CS:
5FF0:0722 A3AC00      MOV    [00AC],AX
5FF0:0725 E86500      CALL   078D
5FF0:0728 E87300      CALL   079E
5FF0:072B E8D000      CALL   07FE
5FF0:072E 2E          CS:
5FF0:072F A1A000      MOV    AX,[00A0]
5FF0:0732 C3          RET
5FF0:0733 FC          CLD
5FF0:0734 32C0        XOR    AL,AL
5FF0:0736 33FF        XOR    DI,DI
5FF0:0738 2E          CS:
5FF0:0739 8E06B200    MOV    ES,[00B2]
5FF0:073D 26          ES:
5FF0:073E 8E062C00    MOV    ES,[002C]
```

```
5FF0:0742 2E            CS:
5FF0:0743 8C06BC00      MOV     [00BC],ES
5FF0:0747 EB05          JMP     074E
5FF0:0749 B9FFFF        MOV     CX,FFFF
5FF0:074C F2            REPNZ
5FF0:074D AE            SCASB
5FF0:074E AE            SCASB
5FF0:074F 75F8          JNZ     0749
5FF0:0751 2E            CS:
5FF0:0752 893EBE00      MOV     [00BE],DI
5FF0:0756 2E            CS:
5FF0:0757 893EC000      MOV     [00C0],DI
5FF0:075B E84D29        CALL    30AB
5FF0:075E 7210          JB      0770
5FF0:0760 7507          JNZ     0769
5FF0:0762 2E            CS:
5FF0:0763 810EA4000140  OR      WORD PTR [00A4],4001
5FF0:0769 2E            CS:
5FF0:076A 810EA4000080  OR      WORD PTR [00A4],8000
5FF0:0770 E83D29        CALL    30B0
5FF0:0773 7217          JB      078C
5FF0:0775 750E          JNZ     0785
5FF0:0777 2E            CS:
5FF0:0778 810EA6000140  OR      WORD PTR [00A6],4001
5FF0:077E 2E            CS:
5FF0:077F 810EA0000100  OR      WORD PTR [00A0],0001
5FF0:0785 2E            CS:
5FF0:0786 810EA6000080  OR      WORD PTR [00A6],8000
5FF0:078C C3            RET
5FF0:078D 54            PUSH    SP
5FF0:078E 58            POP     AX
5FF0:078F 3BC4          CMP     AX,SP
5FF0:0791 B80100        MOV     AX,0001
5FF0:0794 7403          JZ      0799
5FF0:0796 B80200        MOV     AX,0002
5FF0:0799 2E            CS:
5FF0:079A A3A200        MOV     [00A2],AX
5FF0:079D C3            RET
5FF0:079E 2E            CS:
5FF0:079F A1A400        MOV     AX,[00A4]
5FF0:07A2 A90080        TEST    AX,8000
5FF0:07A5 7407          JZ      07AE
5FF0:07A7 A90040        TEST    AX,4000
5FF0:07AA 7533          JNZ     07DF
```

```
5FF0:07AC EB29         JMP     07D7
5FF0:07AE 2E           CS:
5FF0:07AF F706A2000200 TEST    WORD PTR [00A2],0002
5FF0:07B5 7509         JNZ     07C0
5FF0:07B7 CD11         INT     11
5FF0:07B9 250200       AND     AX,0002
5FF0:07BC 7521         JNZ     07DF
5FF0:07BE EB17         JMP     07D7
5FF0:07C0 DBE3                 FINIT
5FF0:07C2 55           PUSH    BP
5FF0:07C3 8BEC         MOV     BP,SP
5FF0:07C5 B8FFFF       MOV     AX,FFFF
5FF0:07C8 50           PUSH    AX
5FF0:07C9 DD7EFE               FSTSW   [BP-02]
5FF0:07CC B91400       MOV     CX,0014
5FF0:07CF E2FE         LOOP    07CF
5FF0:07D1 58           POP     AX
5FF0:07D2 5D           POP     BP
5FF0:07D3 84C0         TEST    AL,AL
5FF0:07D5 7408         JZ      07DF
5FF0:07D7 2E           CS:
5FF0:07D8 C706B0000100 MOV     WORD PTR [00B0],0001
5FF0:07DE C3           RET
5FF0:07DF 2E           CS:
5FF0:07E0 810EA4000100 OR      WORD PTR [00A4],0001
5FF0:07E6 2E           CS:
5FF0:07E7 C706B0000200 MOV     WORD PTR [00B0],0002
5FF0:07ED DBE3                 FINIT
5FF0:07EF 9B           WAIT
5FF0:07F0 2E           CS:
5FF0:07F1 D93EFC00             FSTCW   [00FC]
5FF0:07F5 9B           WAIT
5FF0:07F6 2E           CS:
5FF0:07F7 8126FC0040E0 AND     WORD PTR [00FC],E040
5FF0:07FD C3           RET
5FF0:07FE B40F         MOV     AH,0F
5FF0:0800 CD10         INT     10
5FF0:0802 3C07         CMP     AL,07
5FF0:0804 724C         JB      0852
5FF0:0806 7774         JA      087C
5FF0:0808 2E           CS:
5FF0:0809 810EA8000100 OR      WORD PTR [00A8],0001
5FF0:080F 2E           CS:
5FF0:0810 8126A000FEFF AND     WORD PTR [00A0],FFFE
```

```
5FF0:0816 E89400        CALL    08AD
5FF0:0819 720F          JB      082A
5FF0:081B 80FF01        CMP     BH,01
5FF0:081E 750A          JNZ     082A
5FF0:0820 2E            CS:
5FF0:0821 810EA8000400  OR      WORD PTR [00A8],0004
5FF0:0827 E98200        JMP     08AC
5FF0:082A BABA03        MOV     DX,03BA
5FF0:082D 32DB          XOR     BL,BL
5FF0:082F EC            IN      AL,DX
5FF0:0830 2480          AND     AL,80
5FF0:0832 8AE0          MOV     AH,AL
5FF0:0834 B90080        MOV     CX,8000
5FF0:0837 EC            IN      AL,DX
5FF0:0838 2480          AND     AL,80
5FF0:083A 3AC4          CMP     AL,AH
5FF0:083C 7407          JZ      0845
5FF0:083E FEC3          INC     BL
5FF0:0840 80FB0A        CMP     BL,0A
5FF0:0843 7404          JZ      0849
5FF0:0845 E2F0          LOOP    0837
5FF0:0847 EB63          JMP     08AC
5FF0:0849 2E            CS:
5FF0:084A 810EA8004000  OR      WORD PTR [00A8],0040
5FF0:0850 EB5A          JMP     08AC
5FF0:0852 2E            CS:
5FF0:0853 810EA8000200  OR      WORD PTR [00A8],0002
5FF0:0859 E85100        CALL    08AD
5FF0:085C 731E          JNB     087C
5FF0:085E E86B00        CALL    08CC
5FF0:0861 7336          JNB     0899
5FF0:0863 2E            CS:
5FF0:0864 F706A6000080  TEST    WORD PTR [00A6],8000
5FF0:086A 7540          JNZ     08AC
5FF0:086C 2E            CS:
5FF0:086D 810EA6000100  OR      WORD PTR [00A6],0001
5FF0:0873 2E            CS:
5FF0:0874 810EA0000100  OR      WORD PTR [00A0],0001
5FF0:087A EB30          JMP     08AC
5FF0:087C E82E00        CALL    08AD
5FF0:087F 7218          JB      0899
5FF0:0881 2E            CS:
5FF0:0882 8126A000FEFF  AND     WORD PTR [00A0],FFFE
5FF0:0888 2E            CS:
```

```
5FF0:0889 881EAA00      MOV     [00AA],BL
5FF0:088D B005          MOV     AL,05
5FF0:088F 0AFF          OR      BH,BH
5FF0:0891 7502          JNZ     0895
5FF0:0893 B01A          MOV     AL,1A
5FF0:0895 2E            CS:
5FF0:0896 A2A800        MOV     [00A8],AL
5FF0:0899 E83000        CALL    08CC
5FF0:089C 720E          JB      08AC
5FF0:089E 2E            CS:
5FF0:089F 810EA8002000  OR      WORD PTR [00A8],0020
5FF0:08A5 2E            CS:
5FF0:08A6 8126A000FEFF  AND     WORD PTR [00A0],FFFE
5FF0:08AC C3            RET
5FF0:08AD B80012        MOV     AX,1200
5FF0:08B0 BB10FF        MOV     BX,FF10
5FF0:08B3 B10F          MOV     CL,0F
5FF0:08B5 CD10          INT     10
5FF0:08B7 32E4          XOR     AH,AH
5FF0:08B9 80F90C        CMP     CL,0C
5FF0:08BC 730C          JNB     08CA
5FF0:08BE 80FF01        CMP     BH,01
5FF0:08C1 7707          JA      08CA
5FF0:08C3 80FB03        CMP     BL,03
5FF0:08C6 7702          JA      08CA
5FF0:08C8 F8            CLC
5FF0:08C9 C3            RET
5FF0:08CA F9            STC
5FF0:08CB C3            RET
5FF0:08CC B8001A        MOV     AX,1A00
5FF0:08CF CD10          INT     10
5FF0:08D1 3C1A          CMP     AL,1A
5FF0:08D3 7516          JNZ     08EB
5FF0:08D5 80FB07        CMP     BL,07
5FF0:08D8 740F          JZ      08E9
5FF0:08DA 80FB08        CMP     BL,08
5FF0:08DD 740A          JZ      08E9
5FF0:08DF 80FB0B        CMP     BL,0B
5FF0:08E2 7207          JB      08EB
5FF0:08E4 80FB0C        CMP     BL,0C
5FF0:08E7 7702          JA      08EB
5FF0:08E9 F8            CLC
5FF0:08EA C3            RET
5FF0:08EB F9            STC
```

```
5FF0:08EC C3            RET
5FF0:08ED 0000          ADD     [BX+SI],AL
5FF0:08EF 0000          ADD     [BX+SI],AL
5FF0:08F1 00E8          ADD     AL,CH
5FF0:08F3 4B            DEC     BX
5FF0:08F4 21CF          AND     DI,CX
5FF0:08F6 0BC0          OR      AX,AX
5FF0:08F8 7812          JS      090C
5FF0:08FA 740F          JZ      090B
5FF0:08FC 51            PUSH    CX
5FF0:08FD 8BC8          MOV     CX,AX
5FF0:08FF E83E21        CALL    2A40
5FF0:0902 B80800        MOV     AX,0008
5FF0:0905 E88A21        CALL    2A92
5FF0:0908 E2F5          LOOP    08FF
```

GaAs

DIGITAL SCANNER

Based on the properties demonstrated with the Bar Code Reader, a less sophisticated scanning device can be built from an IRED. This Digital Scanner is capable of operating in the same fashion as a joystick. In other words, the Digital Scanner generates voltage changes that can be received through a properly configured computer port.

The Digital Scanner can use any of the IREDs previously mentioned in this chapter. When choosing an LED, however, make sure that the detector (Q1) is spectrally matched to the wavelength emission of the LED. If this precaution is not observed, the voltage output from the Digital Scanner will lack resolution in its interpretation of the pattern sample.

Construction Notes

Very few parts are needed for building the Digital Scanner (see Fig. 3-15). During the assembly phase of the construction, ensure that the placement of the IRED (D1) focuses its emission on the spot with the greatest degree of sensitivity for the phototransistor (Q1). This arrangement will take several adjustments prior to operation of the Digital Scanner. Ideally, by leaving the leads for D1 and Q1 approximately 4 mm in length, only slight amounts of bending are necessary to fine tune the sensitivity of this project.

There are two controls for adjusting the output of the Digital Scanner. Potentiometer R3 is used for controlling the sensitivity of the op amp (U1). This adjustment should be placed near its middle range during the focusing

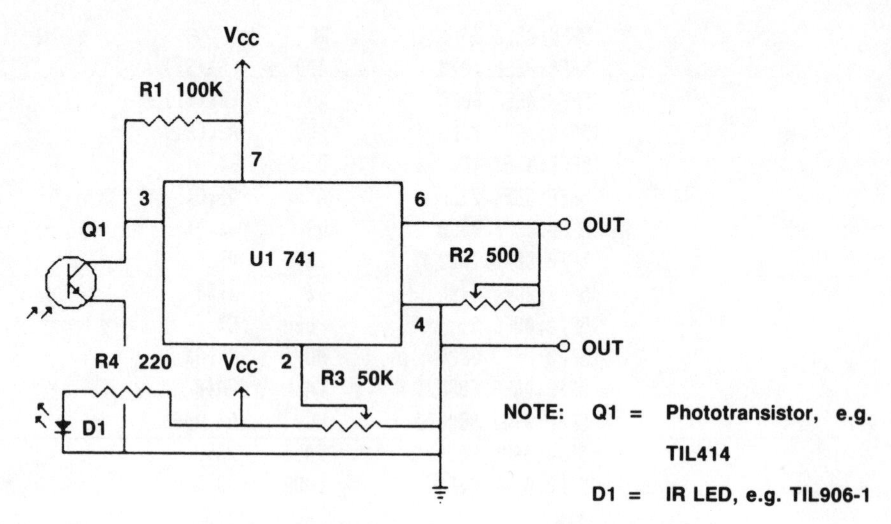

Fig. 3-15. Schematic diagram for Digital Scanner.

of D1 and Q1. Once this configuration has been determined, R3 can be used for manipulating the input of the Digital Scanner for "reading" various types of material.

The other control for adjusting the output of the Digital Scanner is potentiometer R2. Use R2 for changing the output voltage difference prior to the scanner's operation. In other words, R2 should be adjusted for a 0.0 V to + 5.0 V output voltage range during the scanning of a selected pattern. This voltage fluctuation gives the best output for interfacing the Digital Scanner with a voltage-sensitive input port. This type of port is similar to the joystick port found on several personal home microcomputers.

TOUCH-SCREEN DIGITIZER

Interpreting patterns or movements into a computer-digestable form is the function of the digitizer. As outlined in the previous two projects, LEDs make an ideal image sampling sensor for a digitizer. Unfortunately, these digitizers fail to provide the proper degree of resolution for practical use in a graphics-intensive microcomputer system. One failure in obtaining the desired level of resolution is in the focus of the light source. In other words, the tighter the beam of light, the greater the resolving powers of the digitizer.

There are two methods for obtaining high-resolution graphics-oriented digitization. The first improvement is through the incorporation of sophisticated glass lenses into the digitizer's design. Obviously, focusing difficulties and cost overhead prevent the adoption of this resolution-enhancement solution.

A far better method for obtaining a high degree of digitization resolution is by using a memory-mapped digitizing matrix. In effect, this method uses several paired IR emitters/detectors, each of which samples a single bit of data input. Furthermore, when these sensor pairs are arranged in a grid fashion, a complex digitizing matrix can be developed. Essentially, the Touch-Screen Digitizer is an 8X8 matrix implementation of this design.

Construction Notes

The schematic diagram in Fig. 3-16 shows the parts that are needed for building the 8×8 matrix Touch-Screen Digitizer. Only one phototransistor connection has been illustrated in this figure. The remaining 15 phototransistors (Q2-Q16) are interfaced with the NAND gates (U1 and U2). These same phototransistor inputs are also channeled to the twin decoder/multiplexers (U4 and U5) for transmission to a suitable peripheral interface circuit (e.g., Z80-PIO). From this microprocessor interface, the output of the Touch-Screen Digitizer can be directly interpreted by a microcomputer.

WIRELESS MIKE

Voice communications without interconnecting wires are possible through the modulation of an IRED's radiation. These circuits can be connected via fiber optics or assembled in an open-air configuration. This second technique is the form used with this project.

Construction Notes

Two BIFET Quad Op Amps (U1 and U2) are used for the conversion of the voice signals in the Wireless Mike (see Fig. 3-17). These signals are treated differently by both the transmitter and the receiver portions of this circuit.

The transmitter op amp sends the audio input to transistor Q1 for modulation of the IRED D1. The radiation generated by this LED is biased with potentiometer R4 for control over the resolution of the transmitted signal.

In turn, the receiver op amp converts the signal from the phototransistor (Q1) into an output that is capable of driving an external amplifier. If noise becomes a problem in the receiver circuit, a bypass capacitor can be connected between the positive voltage and the circuit ground.

The low parts count for this project makes the Wireless Mike easily translated to a PCB template (see Figs. 3-18 and 3-19). This fabrication method enables the construction of an extremely small transmitting and receiving package.

Operation

When both the IR transmitter and receiver have been assembled (using either a PCB or wire-to-wire soldering), several adjustments should be made for obtaining the highest possible sound quality. These adjustments deal with the calibration of the resolving abilities of potentiometer R4.

- Place the transmitter/receiver two inches apart.
- Send a signal through the transmitter.
- Adjust R4 for the best sound quality and mark this location.

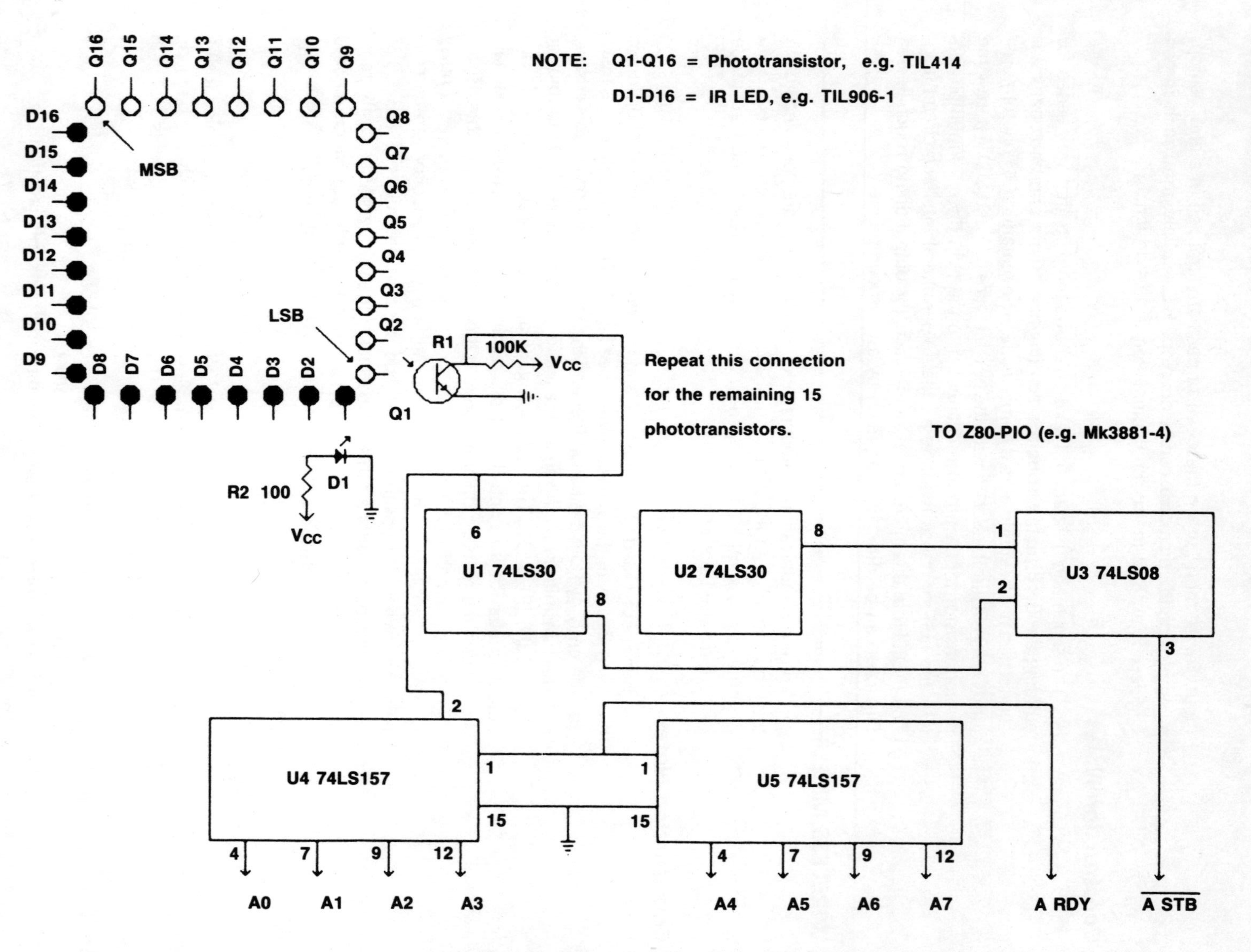

Fig. 3-16. Schematic diagram for Touch-Screen Digitizer.

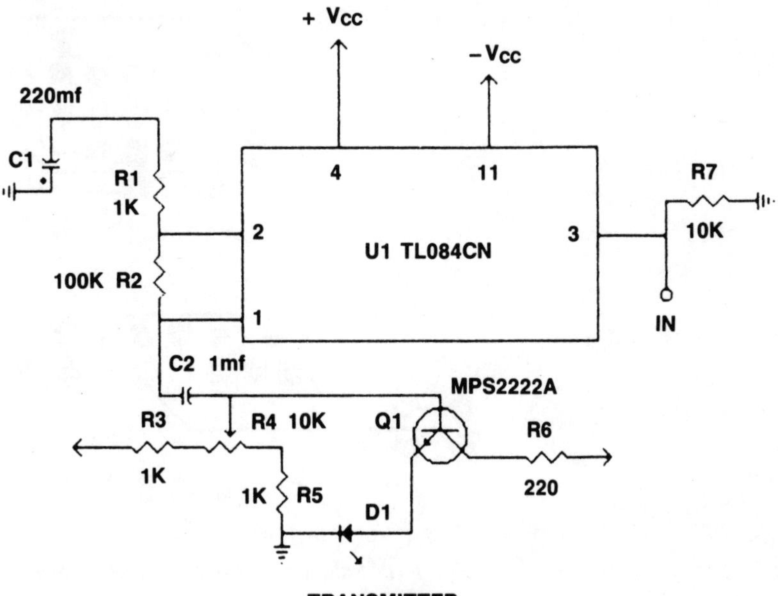

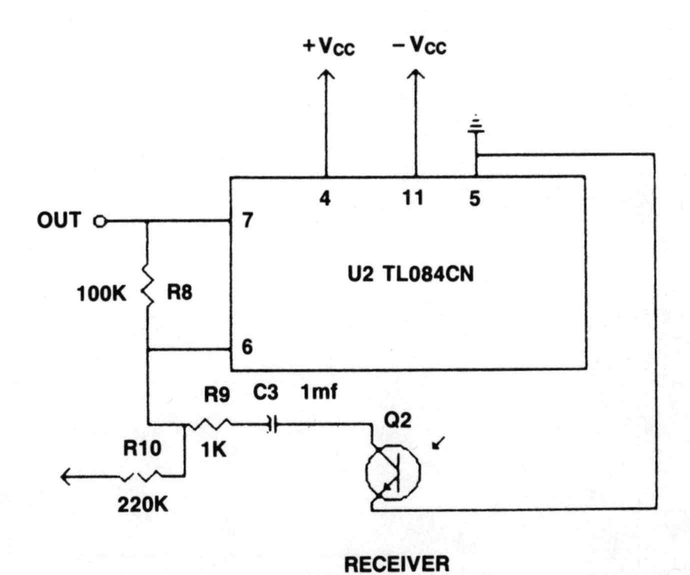

NOTE: D1 - IR LED, e.g. TIL906-1
Q2 - Phototransistor, e.g. TIL414

Fig. 3-17. Schematic diagram for Wireless Mike.

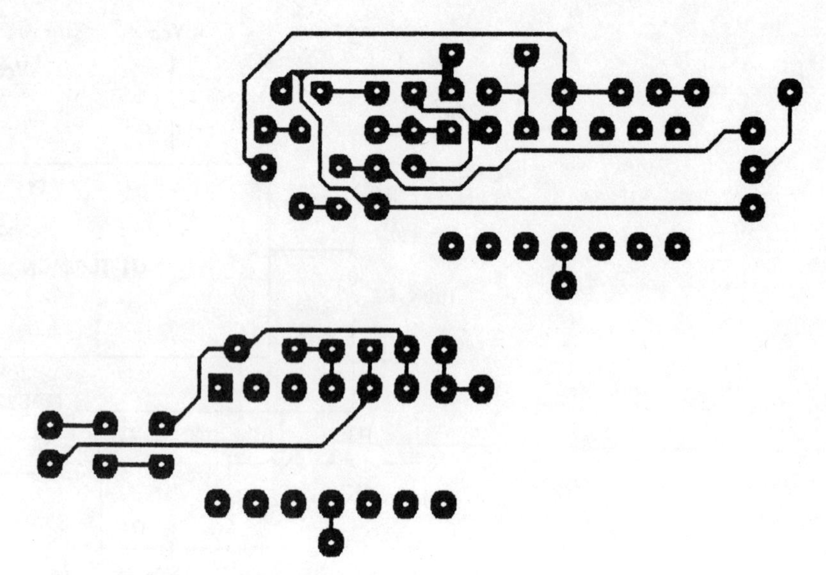

Fig. 3-18. Solder side for the Wireless Mike template, shown at 2X size.

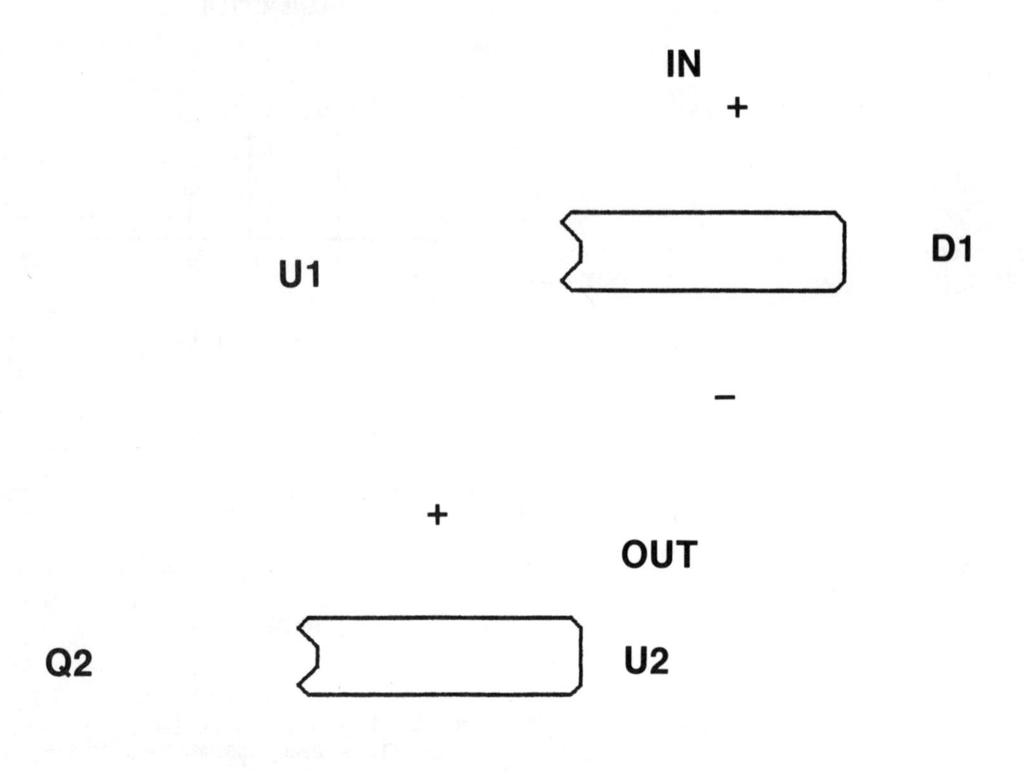

Fig. 3-19. Parts layout for the Wireless Mike PCB.

- Repeat this procedure at the following transmitter-to-receiver distances: 6 inches, 1 foot, 1 yard, 15 feet, and 30 feet. This final test distance might only be obtainable in a darkened room. Greater distances are possible with the Wireless Mike when glass lenses are placed between D1 and Q1.

TACHOMETER

When IREDs with coupled phototransistors are the source of a detection system, various converter ICs can be used for modifying their output. The Tachometer is an example where a single IC is capable of providing wavelength conversion. In this case, a Frequency-to-Voltage Converter (U1 LM2907) IC is used for reading the wavelength input from an LED detection system (see Fig. 3-20).

A chief point to remember when using an LED detection system for reading the reflections from a rotating wheel deals with properly interpreting its spatial frequency. The spatial frequency of a spinning wheel refers to the ratio between dark and light zones or areas of reflectance and transmission. This frequency will then be used for translation into revolutions per time unit. As in all of the previously discussed projects, the alignment of the LED and detector in the Tachometer must be within the angular displacement and the area of sensitivity for these two devices, respectively.

SENSOR

There is a large variety of different photo-reactive devices that can be used for detecting the radiation of an IRED. Prior to this project, phototransistors have been the chief method for receiving the emissions from these light sources. Another detector that is capable of receiving 820 nm wavelengths

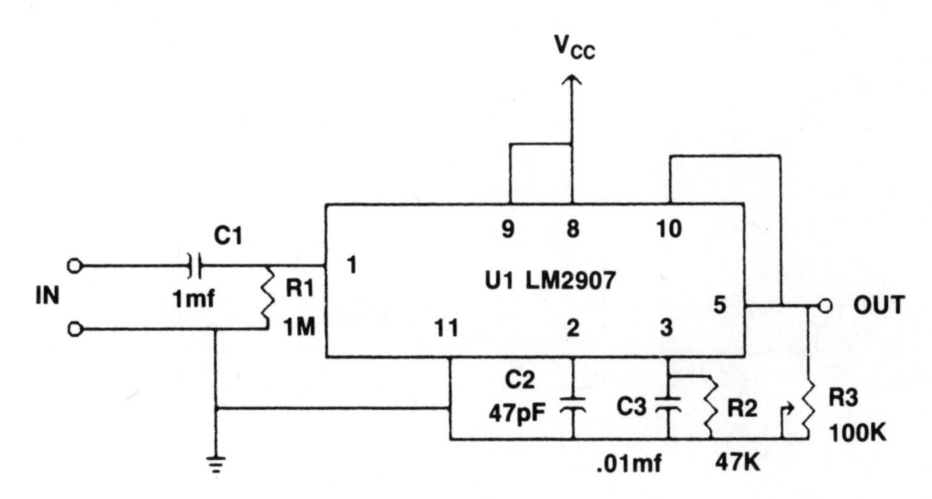

Fig. 3-20. Schematic diagram for Tachometer.

is the Spectronics Schmitt Detector (SD4324). This detector is actually a combination of several circuits enclosed in a single clear housing.

The SD4324 is a TTL or CMOS compatible device holding a photodiode, preamplifier section, Schmitt threshold trigger, and buffer in one 2-pin TO-5 package. The incorporation of a plastic lens into the top of this package provides a field-of-view of approximately 40 degrees.

An extremely simple IR photodetection system can be created from the SD4324 (see Fig. 3-21). The complexity of the Schmitt Detector aids in reducing the parts count for the detector portion of the Sensor to a single invert-

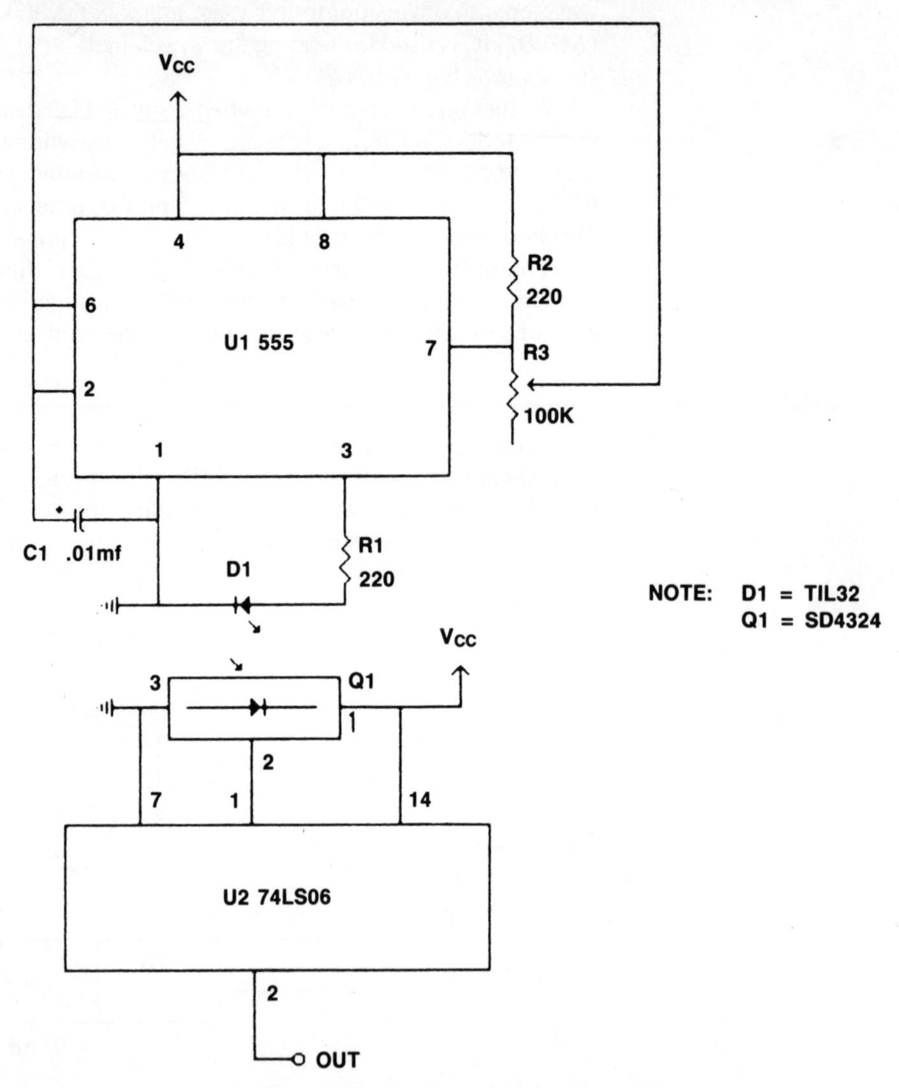

Fig. 3-21. Schematic diagram for Sensor.

This connection method makes the Fiber-Optic Relay's operation independent of atmospheric conditions. For example, the intensity of environmental lighting will not effect or reduce the integrity of this circuit.

The emitter in this project radiates at a wavelength of 820 nm (see Fig. 3-22). This frequency is spectrally matched by the MFOD72 detector. If an IRED is substituted for D1, then Q3 should be altered to match this longer wavelength.

As presented, the Fiber-Optic Relay can only operate with a maximum cable length of 10 meters. By changing the value of resistor R2, this length can be increased to a maximum of 17 meters. Lowering the resistance value of R2 increases the allowable cable length. The ceiling for this resistance reduction is 33 ohms, which generates a forward current of 100 mA in D1.

er/buffer IC (U2). A spectrally-matched IRED (D1) is pulsed by the 555 IC for sending a stream of IR emissions to the SD4324. The effect SD4324 as a sensitive detector can be observed by adding an LED to the port of the 74LS06. This LED will flash at a rate equal to that of the pulsir

FIBER-OPTIC RELAY

One final LED project is the Fiber-Optic Relay. The connection bet the emitter and the detector in this system is via a length of fiber-optic

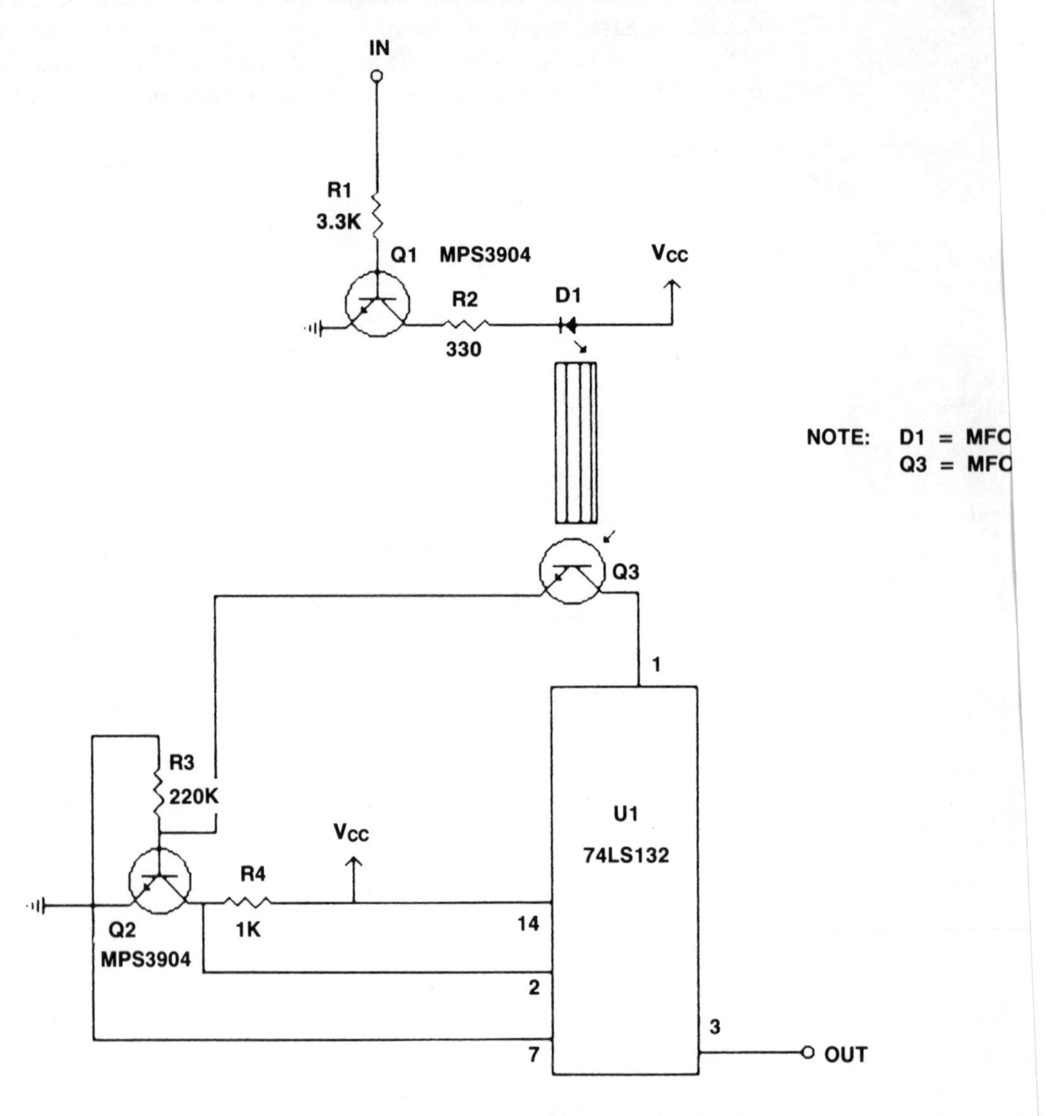

Fig. 3-22. Schematic diagram for Fiber-Optic Relay.

4

Optocouplers

Circuits based on discrete IREDs and detectors can pose several potential stumbling blocks for the designer. First, there is the problem of obtaining IREDs and detectors that are spectrally matched to each other. Second, aligning the emission from the IRED with the detector's area of greatest sensitivity must be within the angular displacement tolerances for each device. Additionally, atmospheric interference must be minimized for ensuring the optimal performance of the detection system. Last and not least, the cost of obtaining matched IR emitter/detector pairs can become prohibitive.

A compact, low-cost answer to all of these design limitations is found in the optocoupler. An optocoupler's typical configuration consists of an IRED and a phototransistor permanently housed inside a sealed DIP carrier (see Fig. 4-1). Even though both of these devices are mounted on the same DIP, they remain electrically isolated from each other with separate and distinct power and ground pins. This isolation is further enhanced by a thin film of optical glass which is sandwiched between the IRED and the phototransistor inside the DIP. Therefore, only a truly optical connection is possible between the IRED and the detector.

By providing only an optical connection between the integral IRED and phototransistor, optocouplers are vitally important in applications where high electrical interference or electromagnetic noise are dominant features. Furthermore, optocouplers are useful as electrical isolators where power surges and output transients could destroy the delicate logic of a CMOS or TTL circuit.

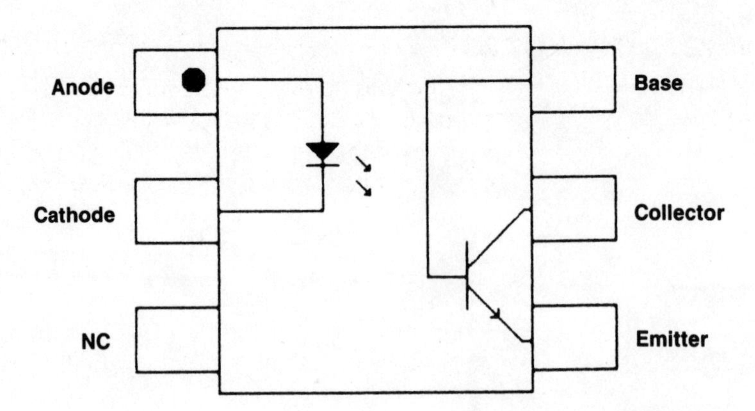

Fig. 4-1. A representative optocoupler's internal configuration.

When designing with optocouplers, there are two performance features that aren't found in discrete IREDs and detectors: *peak isolation voltage* and *current transfer ratio* (CTR). Peak isolation voltage deals with the maximum voltage that can be safely applied to the optocoupler prior to damage or leakage. In this case, damage to the optocoupler results in a short between the input-IRED and the output-phototransistor. This short will circumvent the isolation/insulation benefits that had been previously afforded by the optocoupler.

The current transfer ratio is a statement of how efficiently the optocoupler is capable of passing electricity between the input and the output. The CTR of an optocoupler is calculated as the ratio of the output collector current to the forward LED input current. This ratio is then converted into a percentage at a given forward current.

There are six popular configurations of optocouplers that are widely used in circuit design: phototransistor, photodarlington, split-darlington, logic interface, triac, and SCR (silicon controlled rectifier). Briefly, each of these optocouplers offers the designer a choice in interface control.

PHOTOTRANSISTOR

The most common optocoupler is the phototransistor. Consisting of an emitter and a detector, the phototransistor offers a high-speed interface with a bandwidth in excess of 100 kHz. The greatest trade-off with the phototransistor, however, is a modest output current.

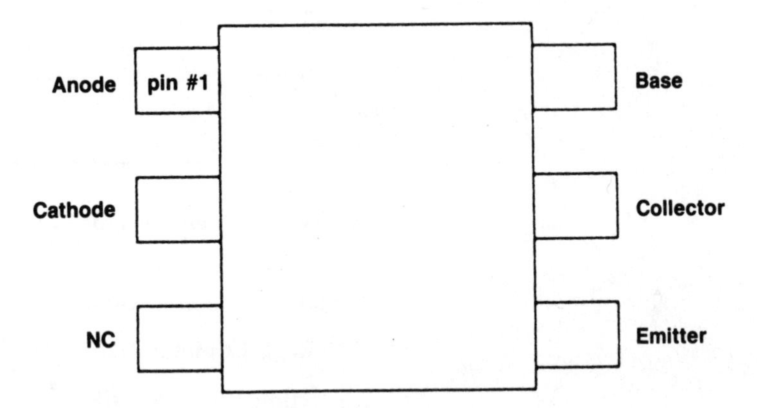

Fig. 4-2. Pin assignments for General Instrument 4N25.

Product Example: General Instrument 4N25

Package Configuration: 6-pin DIP

Isolation Voltage: 2500 V

CTR: 20% with a forward current of 10 mA

Total Power Dissipation: 250 mW

Bandwidth: 300 kHz

Collector Output Current: 5.0 mA

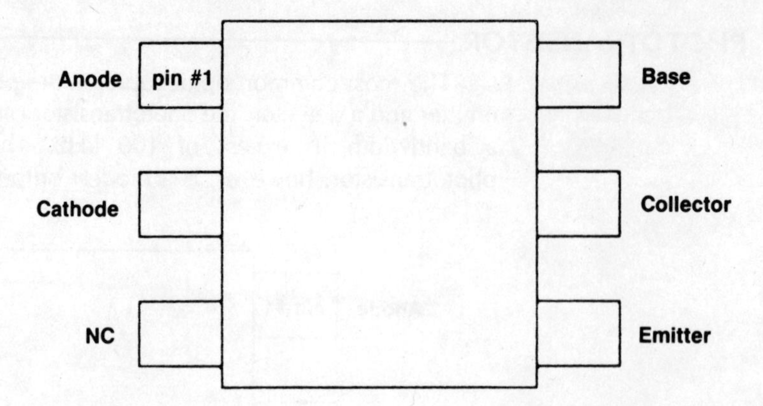

Fig. 4-3. Pin assignments for General Instrument 4N35.

Product Example: General Instrument 4N35

Package Configuration: 6-pin DIP

Isolation Voltage: 2500 V

CTR: 40% with a forward current of 10 mA

Total Power Dissipation: 300 mW

Collector Output Current: 2.0 mA

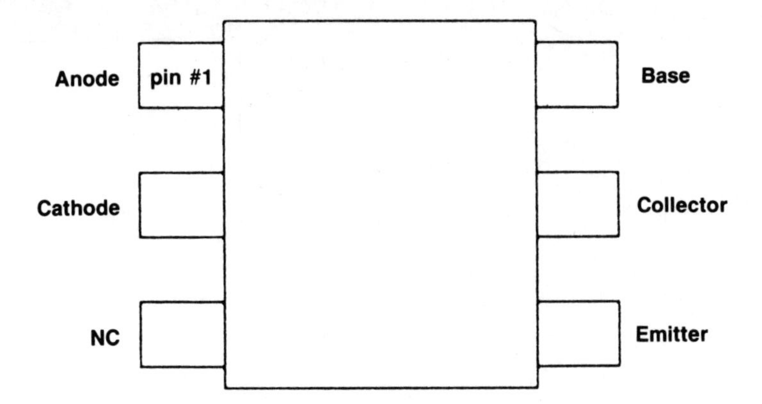

Fig. 4-4. Pin assignments for General Instrument CNX35.

Product Example: General Instrument CNX35

Package Configuration: 6-pin DIP

Isolation Voltage: 4400 V

CTR: 40% with a forward current of 10 mA

Total Power Dissipation: 260 mW

Collector Output Current: 5.0 mA

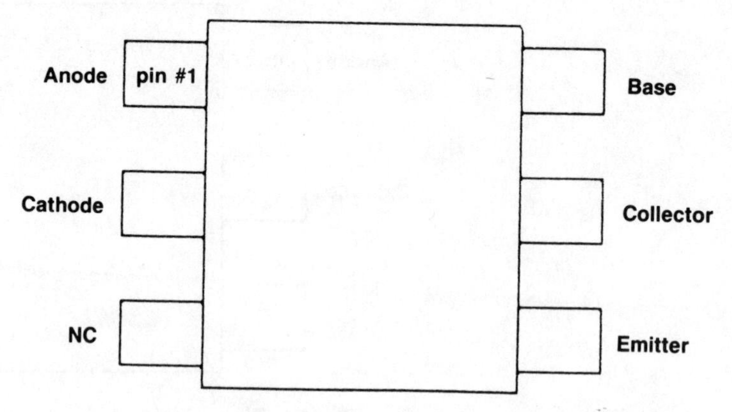

Fig. 4-5. Pin assignments for General Instrument MCT2E.

Product Example: General Instrument MCT2E

Package Configuration: 6-pin DIP

Isolation Voltage: 2500 V

CTR: 60% with a forward current of 10 mA

Total Power Dissipation: 250 mW

Bandwidth: 150 kHz

Collector Output Current: 2.0 mA

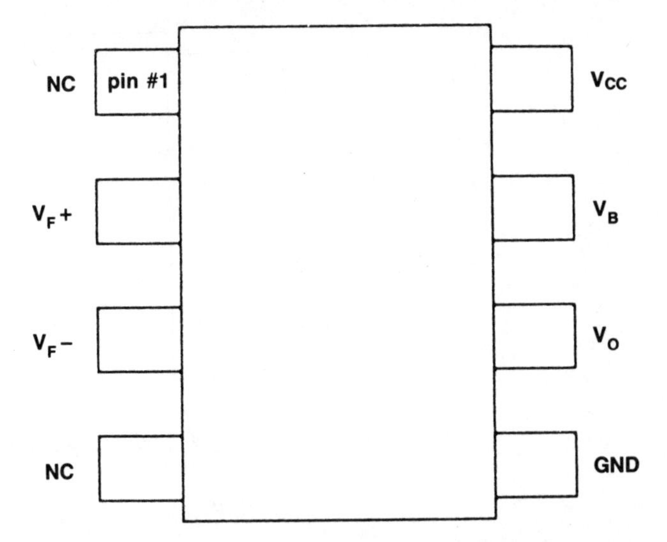

Fig. 4-6. Pin assignments for General Instrument 6N135.

Product Example: General Instrument 6N135

Package Configuration: 8-pin DIP

Isolation Voltage: 2500 V

CTR: 36% with a forward current of 10 mA

Total Power Dissipation: 100 mW

Bandwidth: 2 MHz

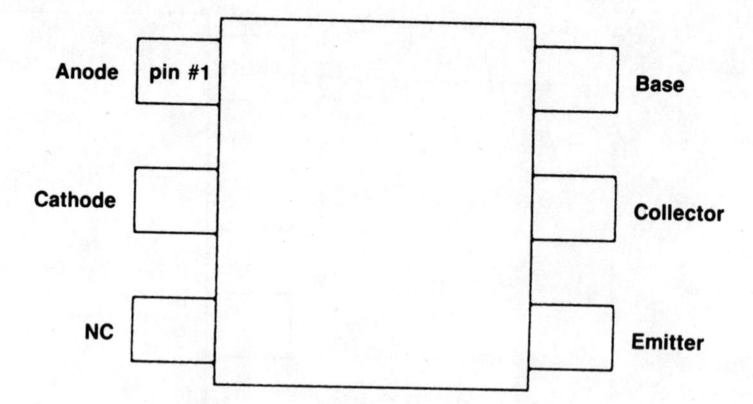

Fig. 4-7. Pin assignments for Motorola TIL112.

Product Example: Motorola TIL112

Package Configuration: 6-pin DIP

Isolation Voltage: 7500 V

CTR: 2% with a forward current of 10 mA

Collector Output Current: 2.0 mA

PHOTODARLINGTON

Offering greater output current at bandwidths under 100 kHz, the photodarlington optocoupler has a CTR greater than the phototransistor.

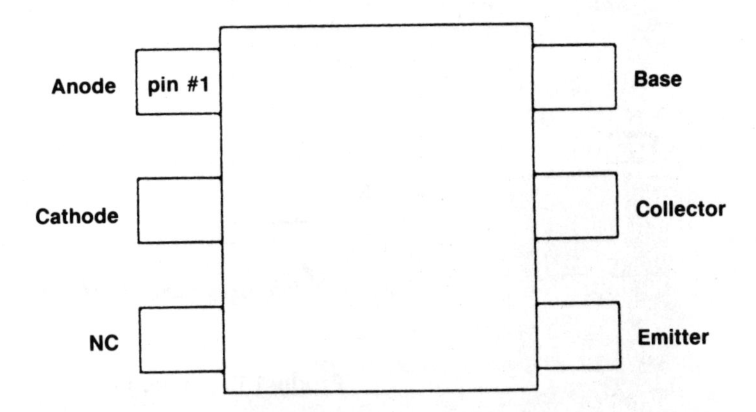

Fig. 4-8. Pin assignments for General Instrument 4N32.

Product Example: General Instrument 4N32

Package Configuration: 6-pin DIP

Isolation Voltage: 2500 V

CTR: 500% with a forward current of 10 mA

Total Power Dissipation: 250 mW

Bandwidth: 30 kHz

Collector Output Current: 50 mA

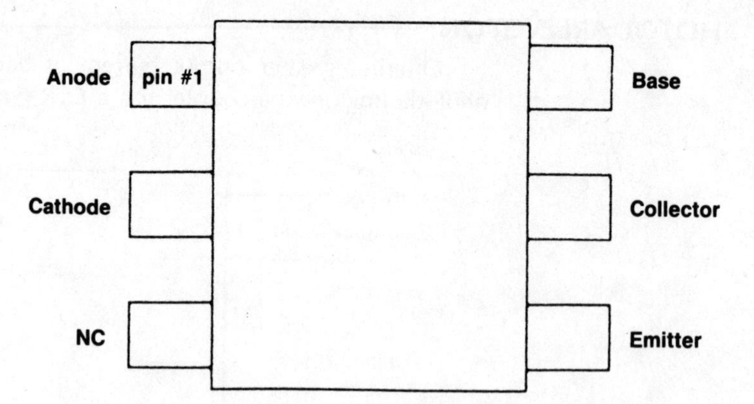

Fig. 4-9. Pin assignments for General Instrument MCA1161.

Product Example: General Instrument MCA1161

Package Configuration: 6-pin DIP

Isolation Voltage: 4000 V

CTR: 1000% with a forward current of 10 mA

Total Power Dissipation: 260 mW

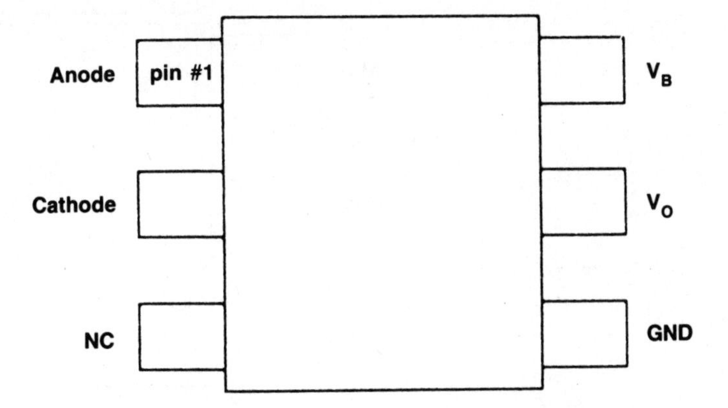

Fig. 4-10. Pin assignments for Hewlett-Packard 4N45.

Product Example: Hewlett-Packard 4N45

Package Configuration: 6-pin DIP

Isolation Voltage: 2500 V

CTR: 500% with a forward current of 10 mA

Total Power Dissipation: 100 mW

Collector Output Current: 60 mA

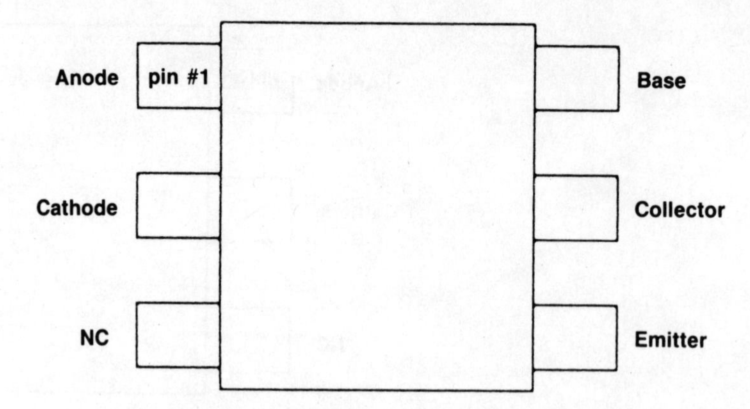

Fig. 4-11. Pin assignments for Motorola H11B1.

Product Example: Motorola H11B1

Package Configuration: 6-pin DIP

Isolation Voltage: 7500 V

CTR: 500% with a forward current of 1.0 mA

Collector Output Current: 1.0 mA

SPLIT-DARLINGTON

An offshoot from the photodarlington optocoupler is the split-darlington. In this version, the emitter is optically coupled to a split-darlington, high-gain detector. This configuration gives the split-darlington both a high CTR and an operation speed advantage over the conventional darlington optocoupler.

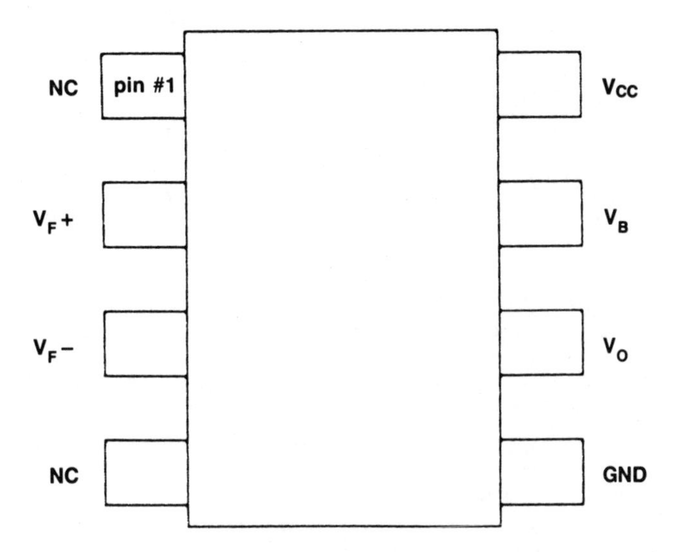

Fig. 4-12. Pin assignments for General Instrument 6N138.

Product Example: General Instrument 6N138

Package Configuration: 8-pin DIP

Isolation Voltage: 2500 V

CTR: 2000% with a forward current of 1.6 mA

Output Power Dissipation: 100 mW

Collector Output Current: 60 mA

LOGIC INTERFACES

These are logic (LSTTL-to-TTL, LSTTL-to-CMOS) gates with no CTR loss and a high isolation voltage. Furthermore, a high operation speed makes these optocouplers ideal for data transmission/reception interfaces that require a baud rate in excess of 1M bps.

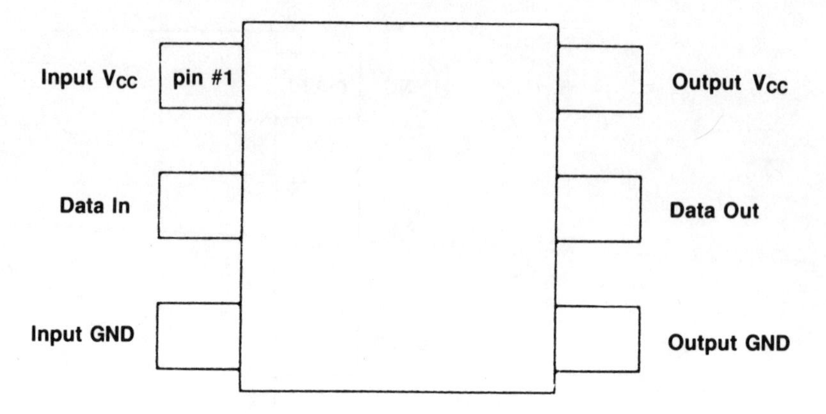

Fig. 4-13. Pin assignments for General Instrument 74OL6010.

Product Example: General Instrument 74OL6010

Package Configuration: 6-pin DIP

Isolation Voltage: 2500 V

Output Power Dissipation: 350 mW

Collector Output Current: 8.0 mA

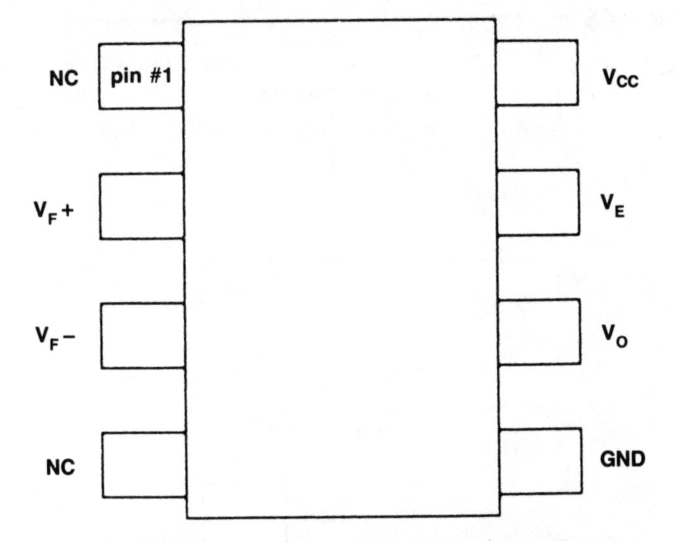

Fig. 4-14. Pin assignments for General Instrument 6N137.

Product Example: General Instrument 6N137

Package Configuration: 8-pin DIP

Isolation Voltage: 2500 V

Output Power Dissipation: 40 mW

TRIACS

A silicon bilateral switch is optically coupled to the IRED for producing isolated triac triggering. This configuration provides low-current ac switching with high isolation voltages.

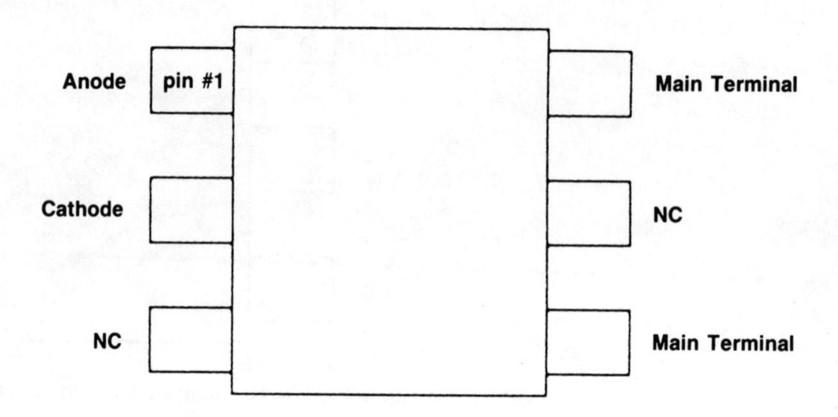

Fig. 4-15. Pin assignments for General Instrument MCP3012.

Product Example: General Instrument MCP3012

Package Configuration: 6-pin DIP

Isolation Voltage: 7500 V

Output Power Dissipation: 300 mW

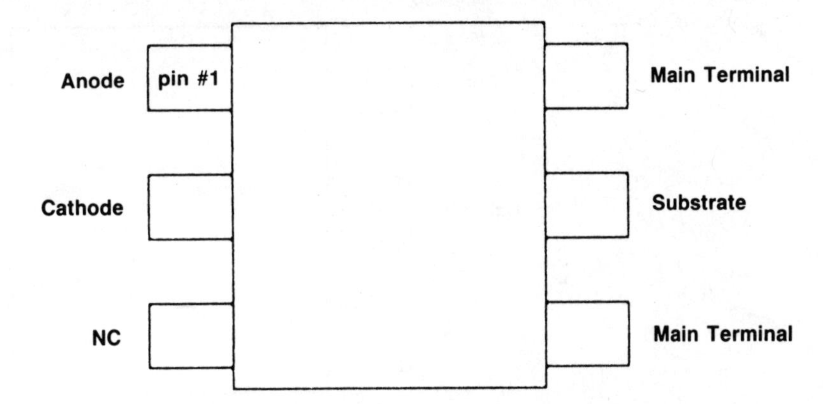

Fig. 4-16. Pin assignments for Motorola H11J1.

Product Example: Motorola H11J1

Package Configuration: 6-pin DIP

Isolation Voltage: 7500 V

Output Power Dissipation: 300 mW

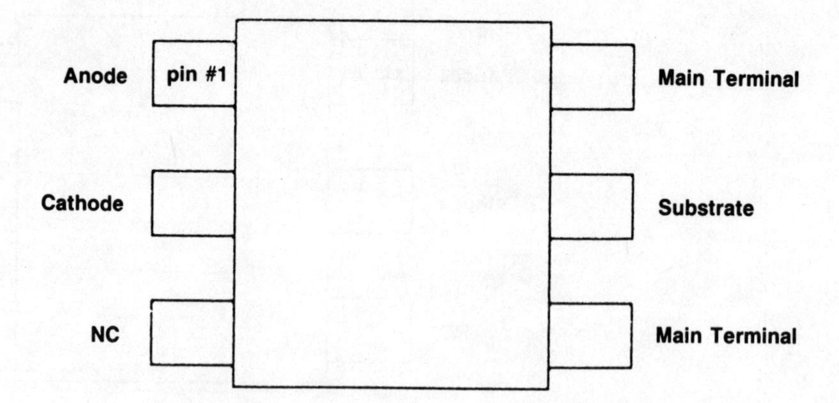

Fig. 4-17. Pin assignments for Motorola MOC3010.

Product Example: Motorola MOC3010

Package Configuration: 6-pin DIP

Isolation Voltage: 7500 V

Output Power Dissipation: 330 mW

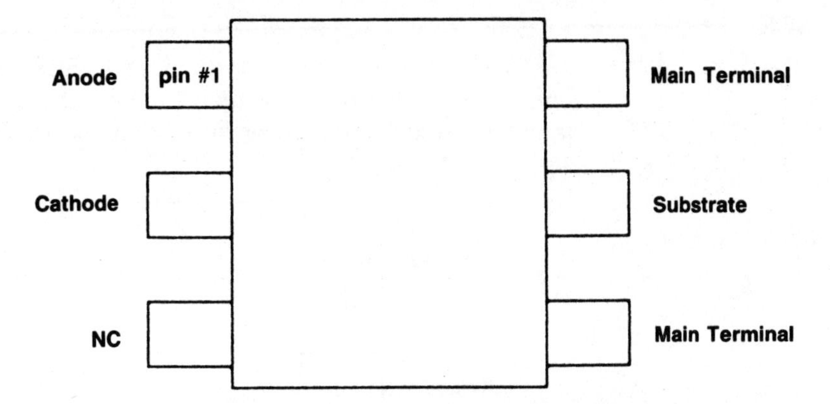

Fig. 4-18. Pin assignments for RCA SK2048.

Product Example: RCA SK2048

Package Configuration: 6-pin DIP T-049

Isolation Voltage: 7500 V

Output Power Dissipation: 330 mW

SCR

High-current-handling capabilities give SCR optocouplers better switching characteristics over triac optocouplers. There is a trade-off, however, in a reduction in output power dissipation and isolation voltage.

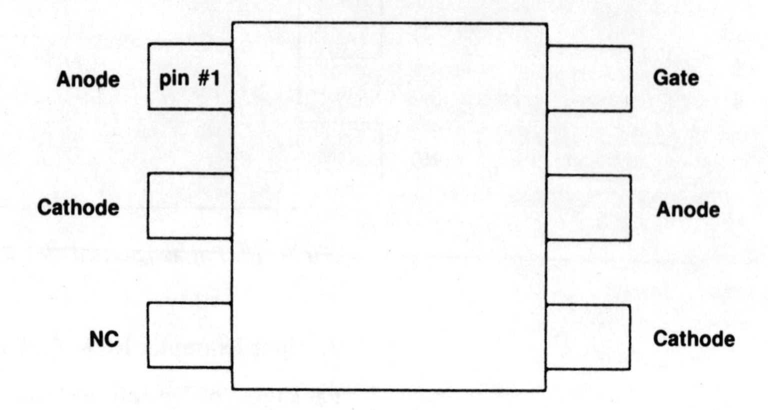

Fig. 4-19. Pin assignments for RCA SK2046.

Product Example: RCA SK2046

Package Configuration: 6-pin DIP T-049

Isolation Voltage: 3550 V

Output Power Dissipation: 250 mW

RS-232C LINE RECEIVER

The high operation speeds of the split-darlington optocoupler make it ideal for use as an RS-232C serial line interface. In this employment, the optocoupler can be used for receiving virtually any range in signal fluctuations and properly converting the output into a TTL level signal.

Construction Notes

Figure 4-20 is a schematic diagram for the RS-232C Line Receiver project. As presented in this circuit, the RS-232C Line Receiver is able to handle baud rates in excess of 35K bps over a 2640-foot line. A limited amount of noise immunity is provided by a hysteresis effect from resistors R2 and R3. If no hysteresis is needed, these resistors may be removed from the circuit.

TELEPHONE LINE MONITOR

Another valuable function performed by the split-darlington optocoupler is voltage isolation. This voltage-filtering feature is particularly valuable in designing telephone interfaces. While a conventional telephone interface doesn't need the sophisticated discrimination afforded by the optocoupler, microcomputer modem designs do need voltage isolation when dealing with telecommunications.

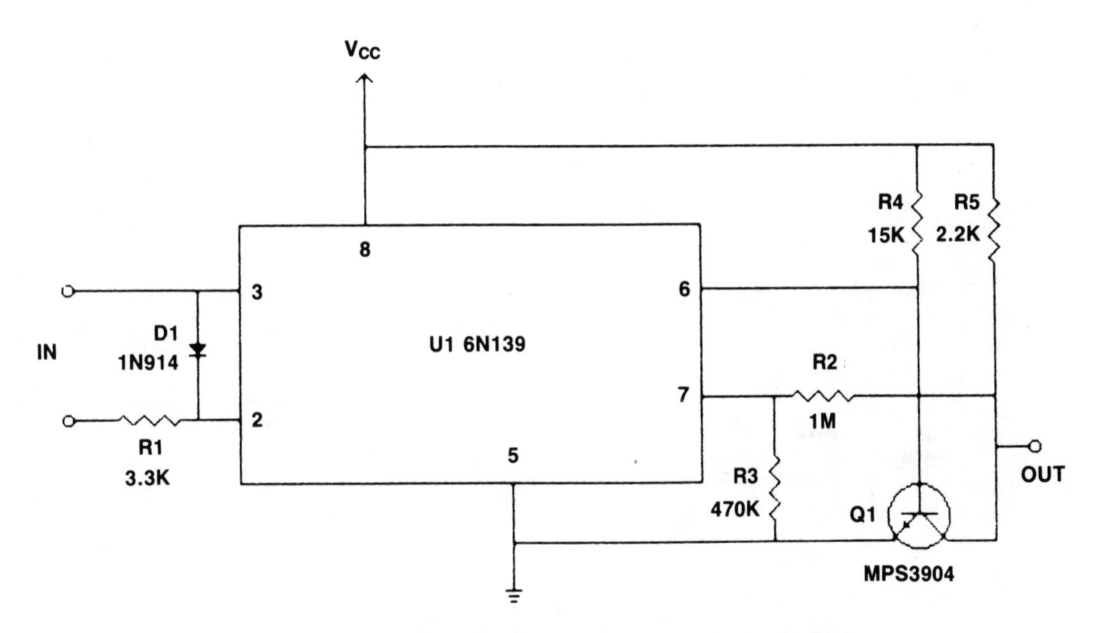

Fig. 4-20. Schematic diagram for RS-232C Line Receiver.

Construction Notes

A modest telephone-to-TTL interface circuit can be built around the 6N139 optocoupler (see Fig. 4-21). This design is insulated from voltage fluctuations up to 2500 V. All telephone line signals can be sensed and interpreted by the Telephone Line Monitor.

GaAs COMPUTER

Interfacing digital devices like microprocessing units (MPUs) to unfiltered power machinery is virtually impossible without making all of the I/O connections through optocouplers. The voltage isolation provided by these connections will then insulate the digital logic from power surges and transients.

Designing a microcomputer based on optocoupler I/O connections can become unwieldy due to the increased cost of parts. One way of reducing this component density is by building a dedicated CPU (central processing unit) that is assigned to performing a particular task. In order to perform this assignment, however, the various logic I/O interfaces that are possible with optocouplers must be understood.

Construction Notes

There are three basic logic circuits that can be easily constructed from optocouplers: TTL-to-TTL, TTL-to-CMOS, and signal inverter. Each of these

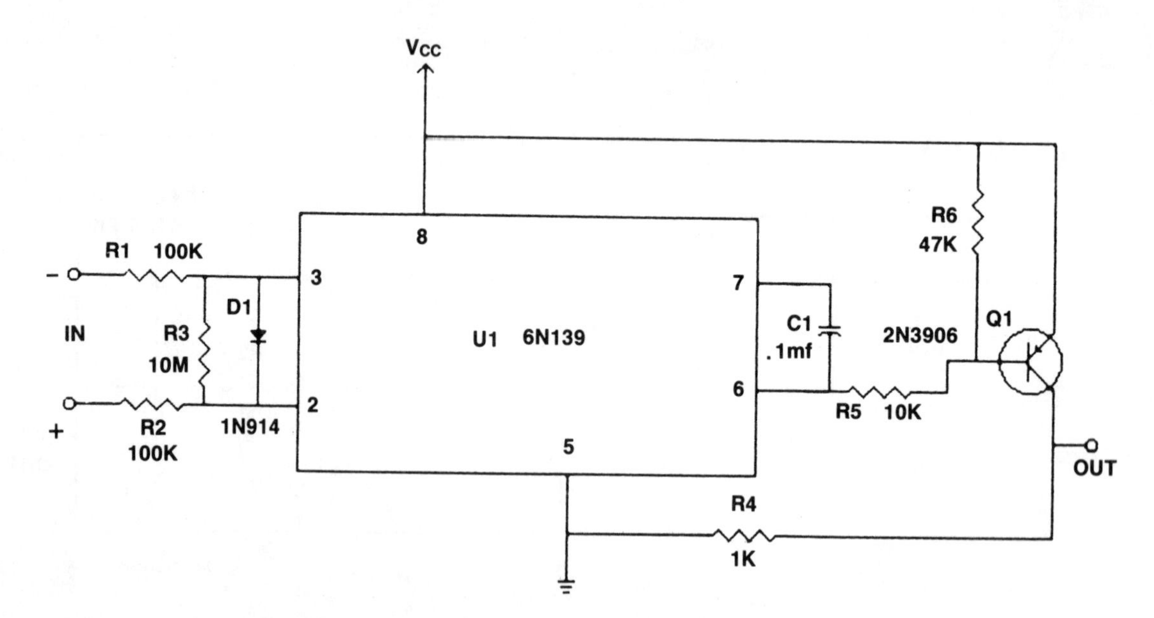

Fig. 4-21. Schematic diagram for Telephone Line Monitor.

circuits can be built with a single optocoupler, a handful of support components, and two TTL (or CMOS) inverters.

The first optocoupler logic circuit is the TTL-to-TTL interface (see Fig. 4-22). The principal optocoupler used in this circuit is the 6N137. Two inverters are used for configuring the input to and the output from the optocoupler. The result is a circuit that is able to operate at speeds of 10 million bits per second (M bps). A 0.1 μF bypass capacitor increases the performance of this circuit.

Another easily implemented logic circuit is the TTL-to-CMOS interface (see Fig. 4-23). Although very similar to the logic circuit in Figure 4-22, this interface uses a CMOS inverter on the optocoupler's output. Therefore, this circuit can interface as well as isolate the different voltage requirements between TTL and CMOS circuitry.

The final practical logic circuit is the signal inverter (see Fig. 4-24). Either TTL or CMOS signals can be inverted with this circuit provided the correct inverter logic is applied to the optocoupler's input and output.

In order to implement these optocoupler logic circuits as control I/O connections, a suitable CPU must be built. Through prudent design, this circuit could be created from as little as five ICs (see Fig. 4-25). The Z-80 MPU was selected as this CPU example based on its processing power versus cost ratio. Therefore, a complete voltage-isolated digital control unit could be built for under $100.

Another important aspect of this Z-80-based CPU design is its ability to function as an excellent optocoupler trainer. In other words, several

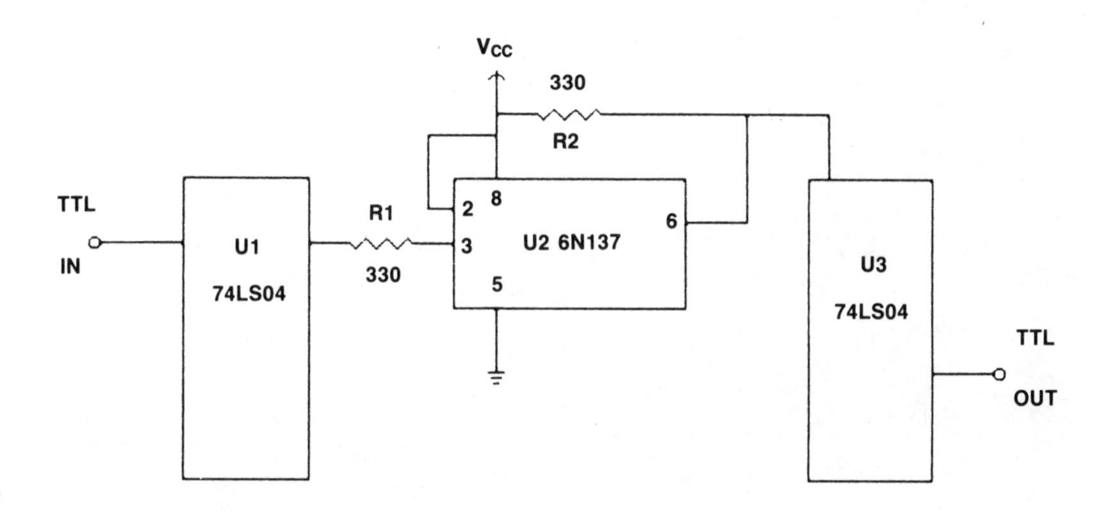

NOTE: Use bypass capacitor for reduced noise.

Fig. 4-22. Schematic diagram for TTL-to-TTL interface.

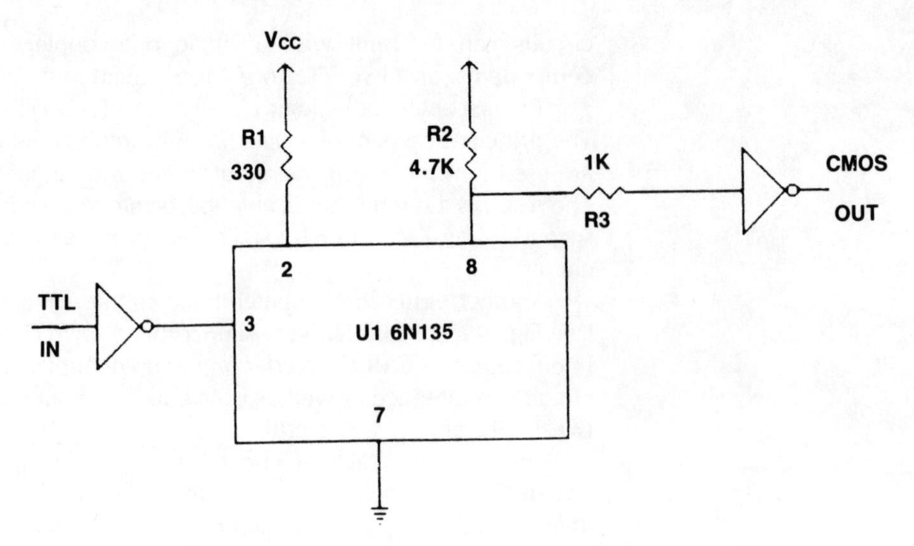

Fig. 4-23. Schematic diagram for TTL-to-CMOS interface.

optocoupler logic circuits could be inexpensively interfaced with the Z-80 PPI (e.g., 8255) and tested in controlled voltage fluctuation experiments. An example of such a trainer is presented as a PCB template (both the solder side and the component side) and parts layout in Figs. 4-26, 4-27, and 4-28, respectively.

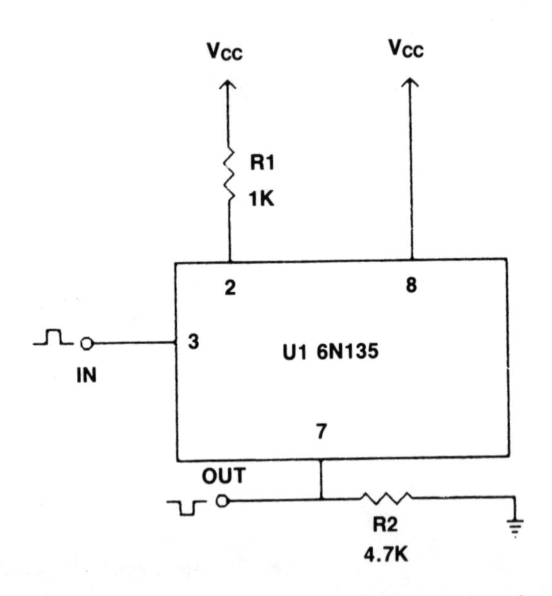

Fig. 4-24. Schematic diagram for signal inverter interface.

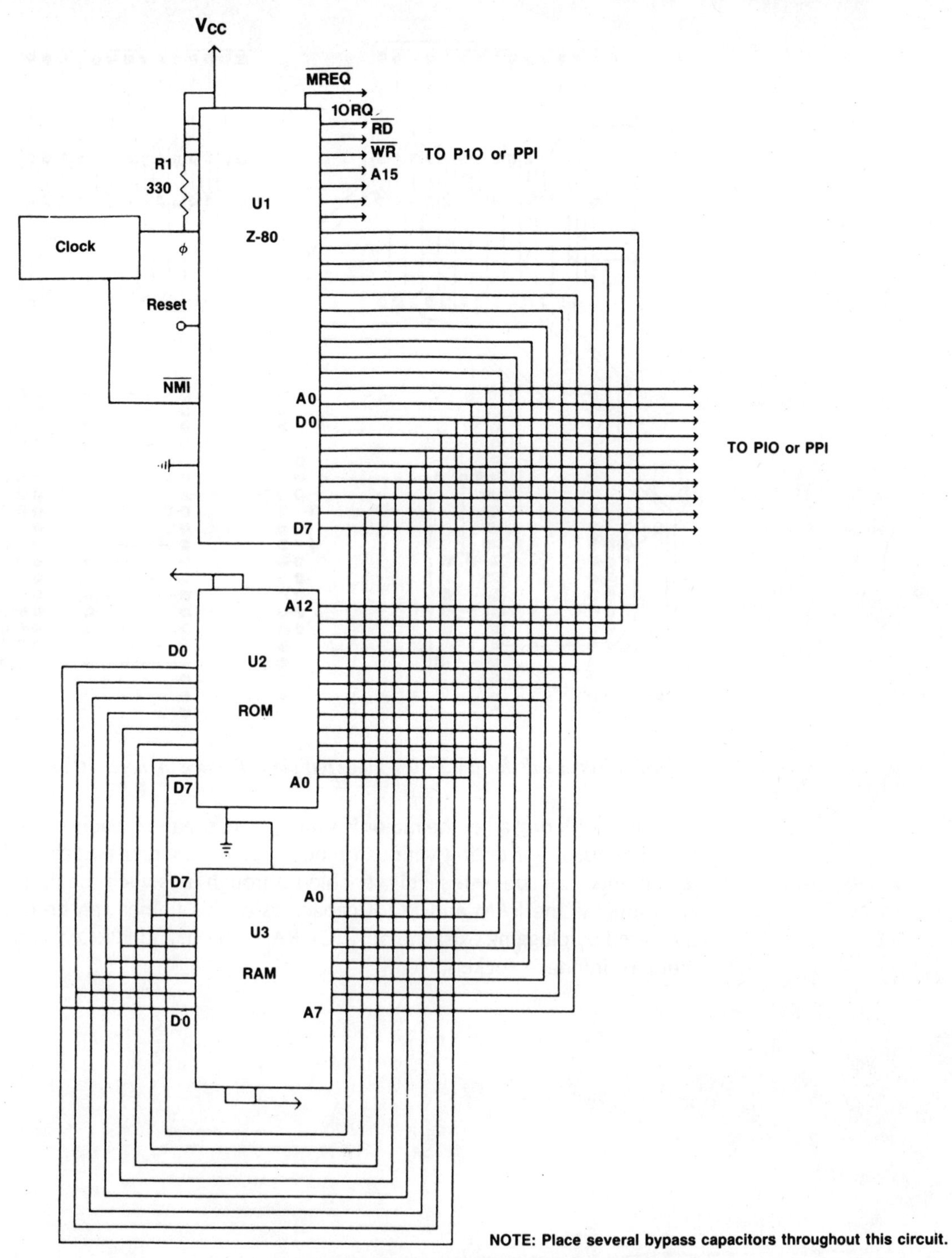

Fig. 4-25. Schematic diagram for Z-80 Trainer.

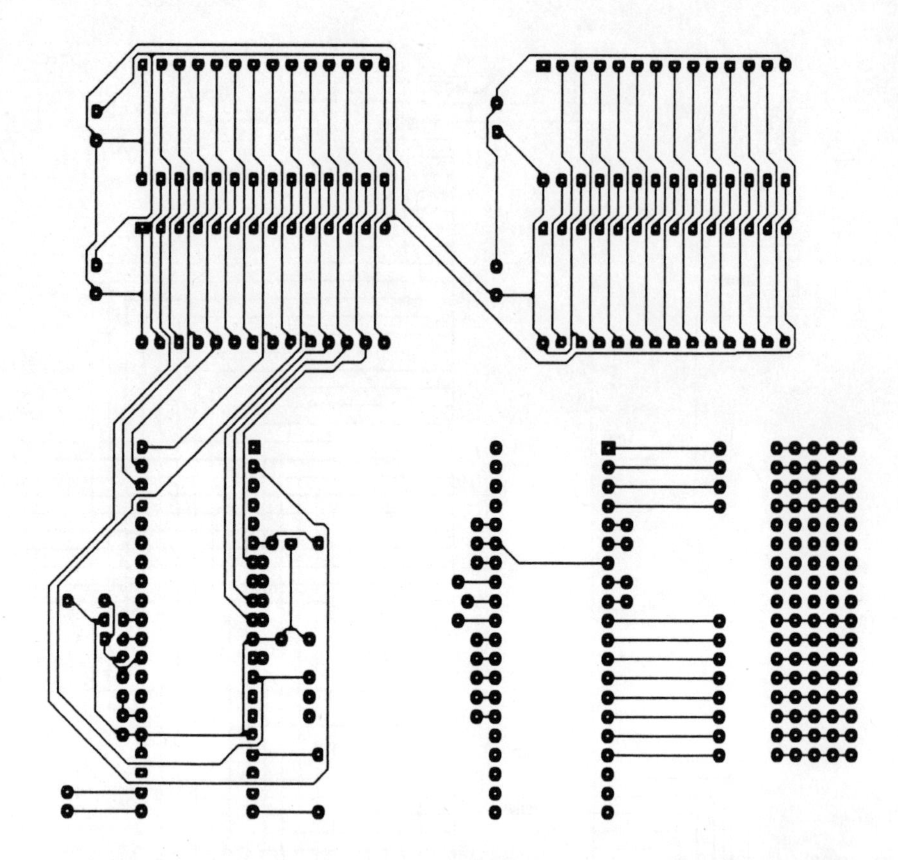

Fig. 4-26. Solder side for the Z-80 Trainer template. Reduce to match IC pin spacing.

After soldering all of the major ICs into place, use the breadboard region of the template for testing various optocoupler logic circuits. Each of these test points can be selectively enabled through a switch bank (SW1). Additionally, the RAM and ROM capacities of this trainer can be quickly increased by plugging extra memory ICs (RAM, ROM, and EPROMs) into the memory interface sockets.

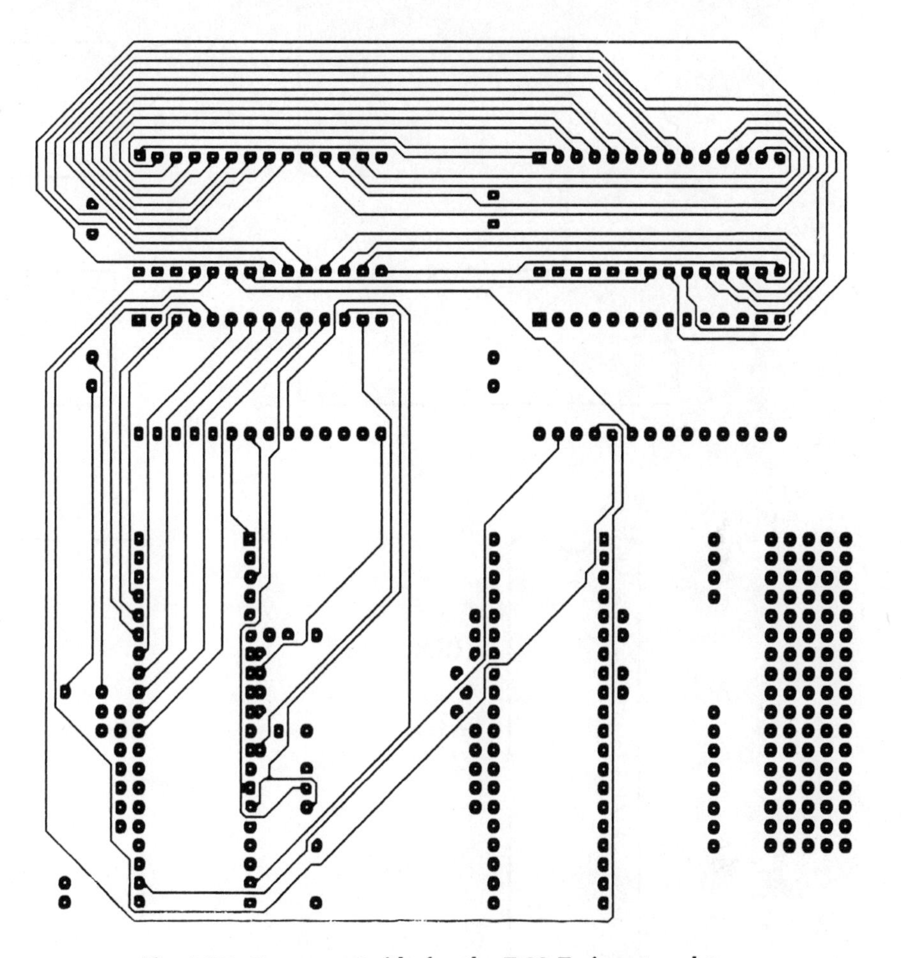

Fig. 4-27. Component side for the Z-80 Trainer template.

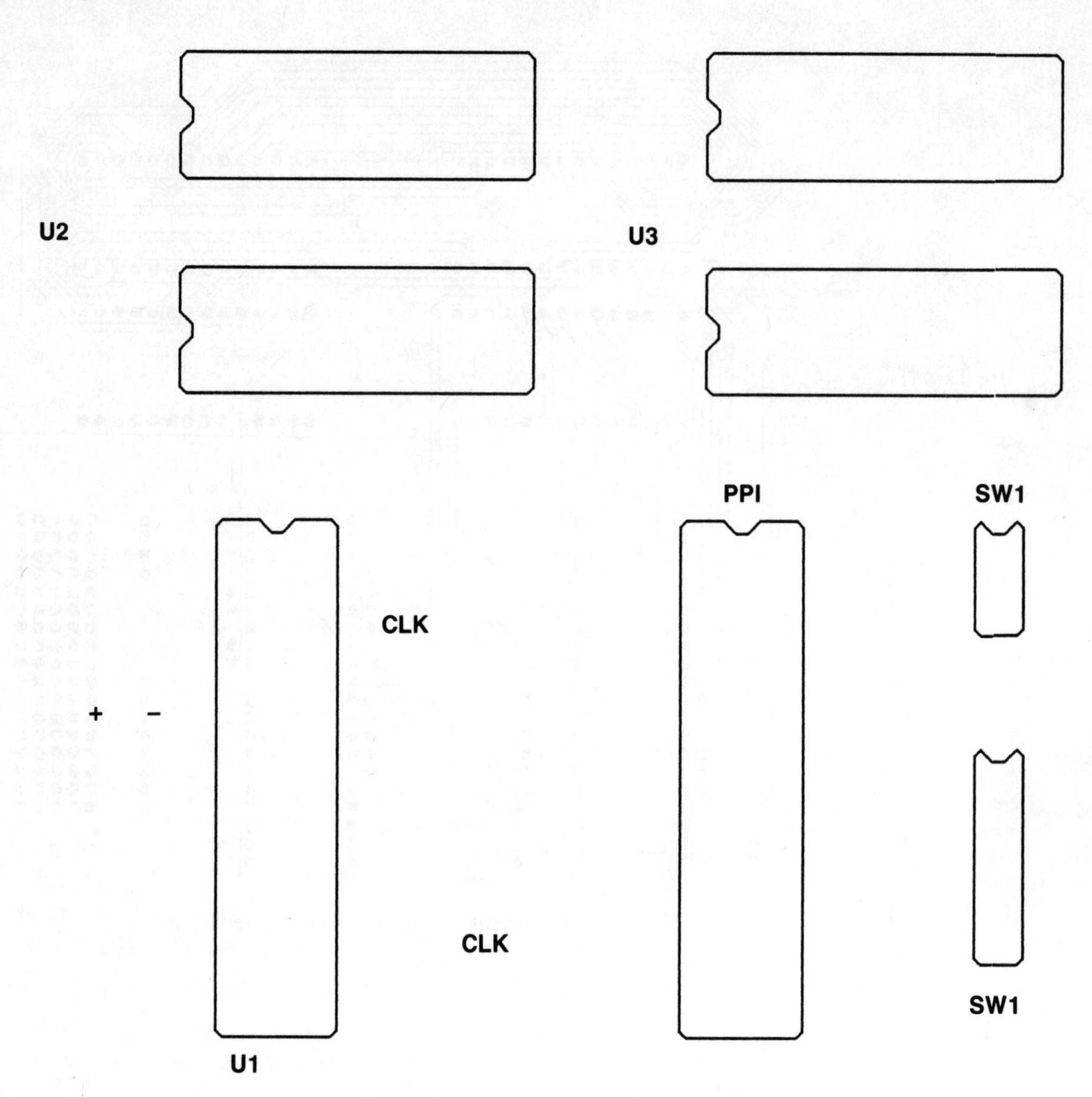

Fig. 4-28. Parts layout for the Z-80 Trainer PCB.

5

Discrete LEDs

Discrete LEDs emit radiations in the visible spectrum that are shorter than infrared and longer than ultraviolet. At these wavelengths, the photons generated by LEDs are visible to the human eye. Typically, there are four colors that can be emitted by LEDs: red (GaAlAs or GaAsP), green (GaP), yellow (GaAsP), and amber (GaAsP). Each of these colors is represented by several different product examples; all of which display different performance specifications. In light of these performance variations, several relationships must be maintained when designing with LEDs.

Chief among the performance specifications to observe in LED applications are the *forward current* and the *reverse voltage*. Ignoring these two power parameters can result in the destruction of the LED.

Controlling the forward current that is applied to an LED is obtained by forward biasing the LED with a resistor (see Fig. 5-1). This resistor must be in series with the LED and the power supply. Determining the correct value for this resistor is a two-step process. First, the voltage drop over the resistor must be established according to the performance specifications for the target LED. Second, after this voltage has been calculated, the actual resistance for the resistor can be determined. Two equations aid you in calculating the final resistor value:

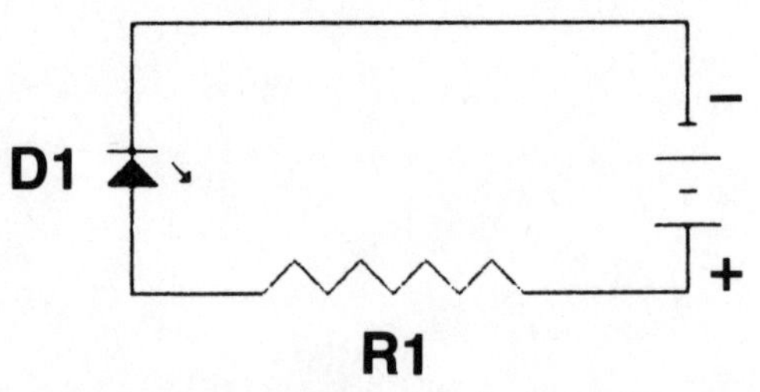

Fig. 5-1. Connecting an LED to a power supply through a current-limiting resistor.

Equation #1: Resistor Voltage Drop

$$Vd = V - Vl$$

where,

Vd = resistor voltage drop
V = power supply in volts
Vl = LED forward voltage drop

Equation #2: Resistor Value

$$R = V/I$$

where,

R = resistor value in ohms
V = resistor voltage drop in volts
I = forward current in amps

Note: The forward current is a value selected by the circuit designer based on the performance specifications for the target LED.

DETERMINING LED FORWARD CURRENT

Purpose: To determine the forward current for a typical LED.

Materials: 200-ohm potentiometer
5V power supply
red LED
multimeter

Procedure:

- Assemble the circuit in Fig. 5-2
- While adjusting the potentiometer, use the multimeter to take volt and amp readings at the indicated points.
- Record these readings.

Results: Compare your results with the specifications for the target LED.

The second major performance specification to observe when designing with LEDs is reverse voltage. The reverse voltage of an LED is expressed in terms of the maximum allowable voltage that can pass through a reverse-biased LED (see Fig. 5-3). There are two solutions to limiting the amount of reverse

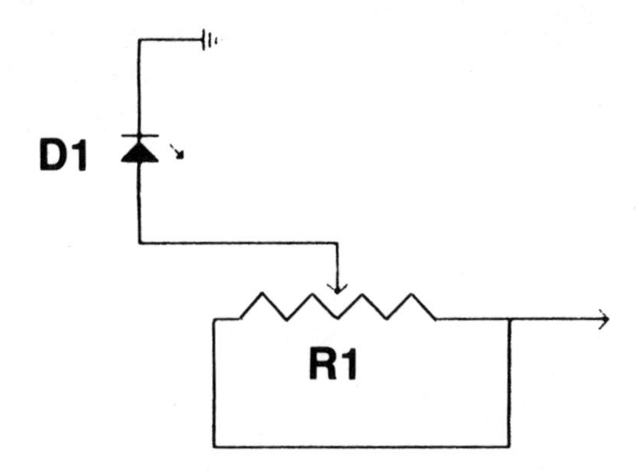

Fig. 5-2. LED forward current test circuit.

voltage applied to an LED. First, a diode can be inverted and installed in parallel to the target LED. This diode will then limit the LED's reverse voltage to the diode's forward voltage drop.

Another method for controlling the reverse voltage of an LED is to replace the inverted diode with an inverted LED. The virtue of this reverse voltage limitation technique is demonstrated with bipolar, bi-color LEDs (e.g., MV5491A). These LEDs report any circuit variations in polarity. This application is useful in ac voltage circuits. When using these bi-color LEDs, a different equation must be used for calculating the series resistor value.

Equation #3: Reverse Voltage Bi-Color LED Resistor Value

$$R = (V - Vl)/I$$

where,

R = resistor value in ohms
V = power voltage in volts
Vl = LED forward voltage in volts
I = forward current in amps

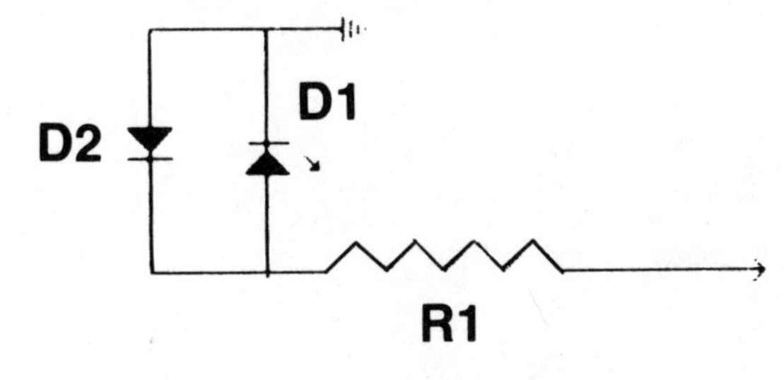

Fig. 5-3. A reverse-biased LED circuit.

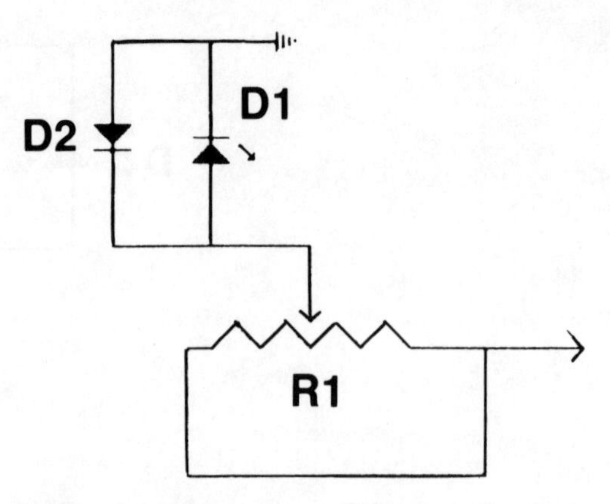

Fig. 5-4. LED reverse voltage test circuit.

Note: The forward current is a value selected by the circuit designer based on the performance specifications for the target LED.

DETERMINING LED REVERSE VOLTAGE

Purpose: To determine the reverse voltage for a typical LED.

Materials: 200-ohm potentiometer
5-V power supply
bi-color LED
multimeter

Procedure:

- Assemble the circuit in Fig. 5-4.
- While adjusting the potentiometer, use the multimeter to take volt and amp readings at the indicated points.
- Record these readings.

Results: Compare your results with the specifications for the target LED.

One last discrete LED that is entering the optoelectronic market is the tri-color LED. In this configuration, three voltages are applied to the package for generating the three different colors: positive dc, negative dc, and ac. Typically, red, green, and amber colors are generated in tri-color LEDs.

DISCRETE LEDs

LEDs can be obtained in four different colors (red, green, yellow, and amber), six different sizes (T-3/4, T-1, T-100, TO-18, T-1 3/4, and rectangular), nine different housing configurations (low profile, clear, diffused, taper, bullet, standoff, narrow angle, wide angle, and maximum contrast white), and three different specialty designs (three-terminal, blinking, and bi-color).

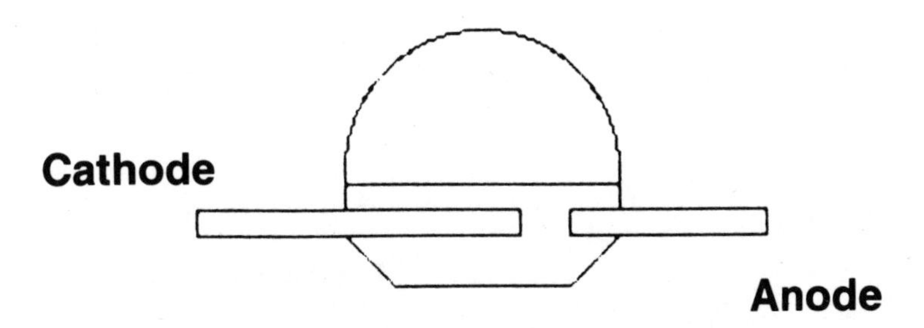

Fig. 5-5. Pin assignments for General Instrument MV50B.

Product Example: General Instrument MV50B

Package Configuration: Subminiature T-3/4

Forward Current: 50 mA

Reverse Voltage: 5.0 V

Wavelength: 660 nm

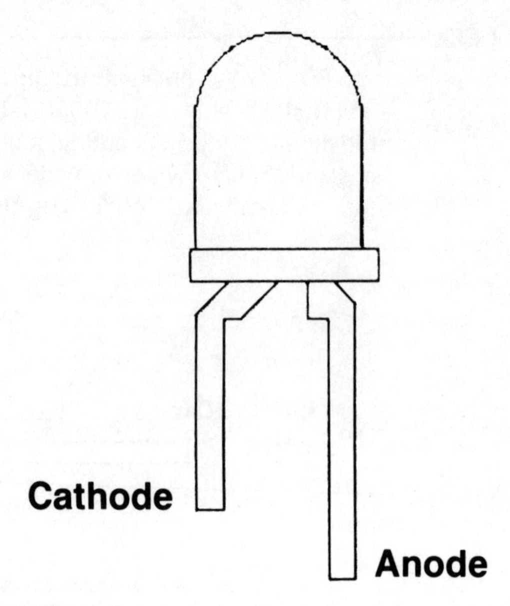

Fig. 5-6. Pin assignments for General Instrument MV5362X.

Product Example: General Instrument MV5362X

Package Configuration: Clear Lens T-100

Forward Current: 30 mA

Reverse Voltage: 5.0 V

Wavelength: 585 nm

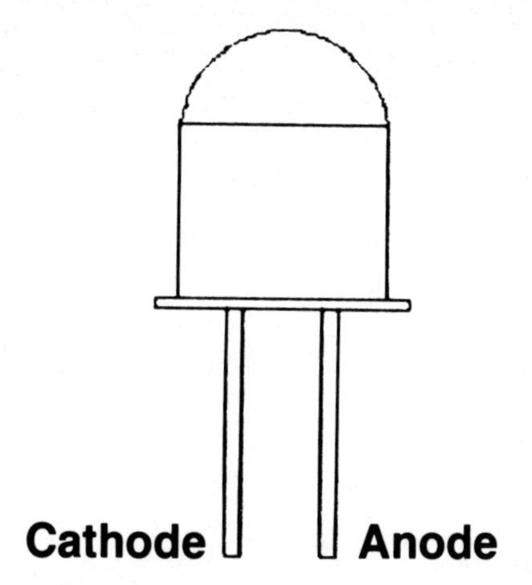

Fig. 5-7. Pin assignments for General Instrument MV10B.

Product Example: General Instrument MV10B

Package Configuration: TO-18

Forward Current: 70 mA

Reverse Voltage: 5.0 V

Wavelength: 660 nm

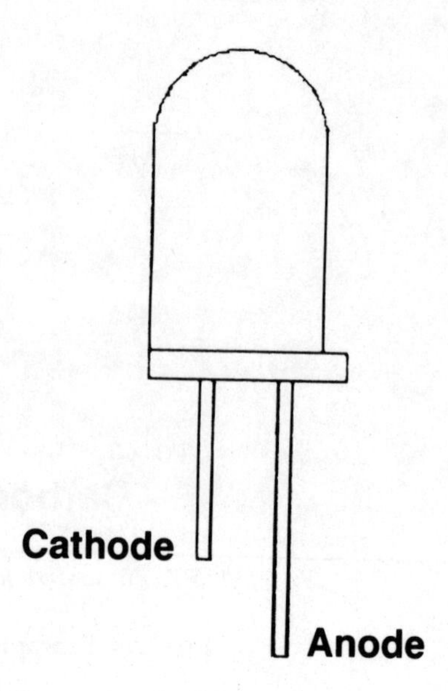

Fig. 5-8. Pin assignments for Hewlett-Packard HLMP-3507.

Product Example: Hewlett-Packard HLMP-3507

Package Configuration: High-performance diffused wide-angle green T-1 3/4

Forward Current: 90 mA

Reverse Voltage: 5.0 V

Wavelength: 565 nm

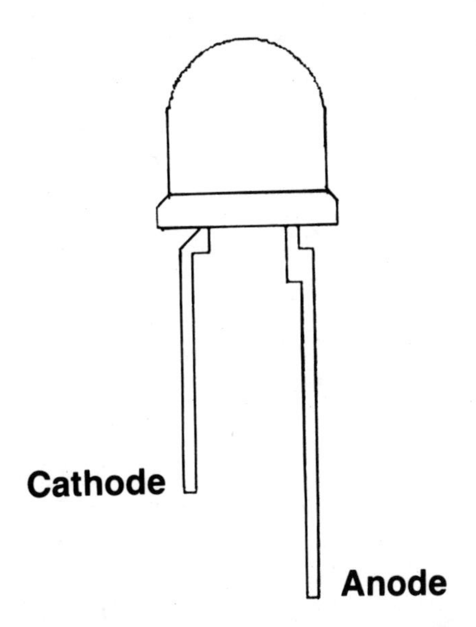

Fig. 5-9. Pin assignments for Hewlett-Packard HLMP-1450.

Product Example: Hewlett-Packard HLMP-1450

Package Configuration: Diffused tinted wide-angle yellow T-1

Forward Current: 60 mA

Reverse Voltage: 5.0 V

Wavelength: 585 nm

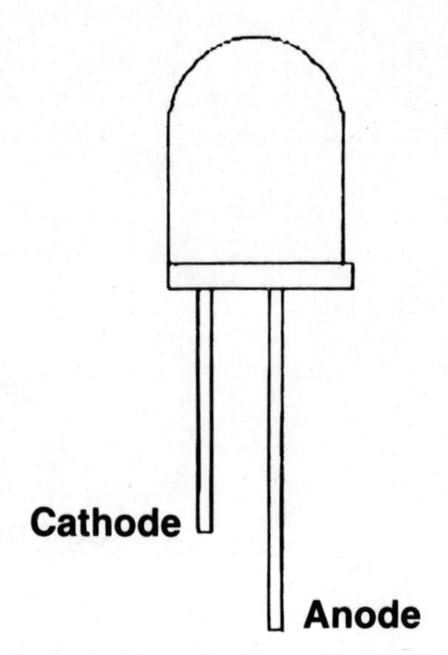

Fig. 5-10. Pin assignments for F336GD.

Product Example: F336GD

Package Configuration: Blinking diffused green T-1 3/4

Forward Current: 35 mA

Reverse Voltage: 0.4 V

Wavelength: 565 nm

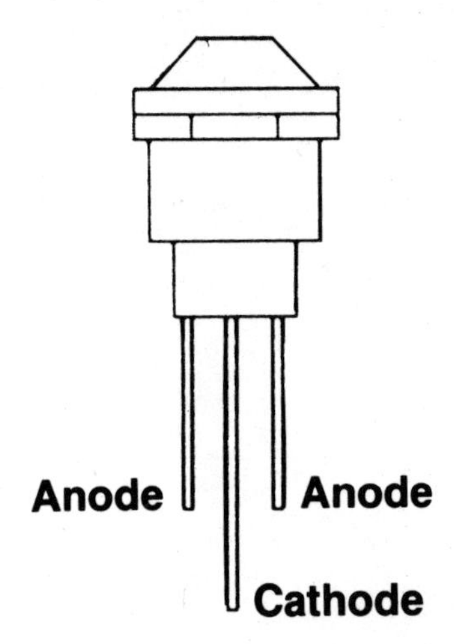

Fig. 5-11. Pin assignments for R9-56.

Product Example: R9-56

Package Configuration: Three-terminal red/green T-1 3/4

Forward Current: 30 mA

Reverse Voltage: 3.0 V

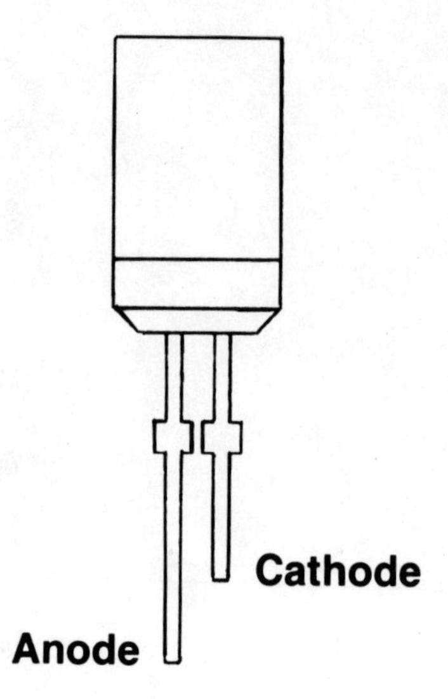

Fig. 5-12. Pin assignments for RCA SK2166.

Product Example: RCA SK2166

Package Configuration: Rectangular red L-012

Forward Current: 30 mA

Reverse Voltage: 3.0 V

Bandwidth: 700 nm

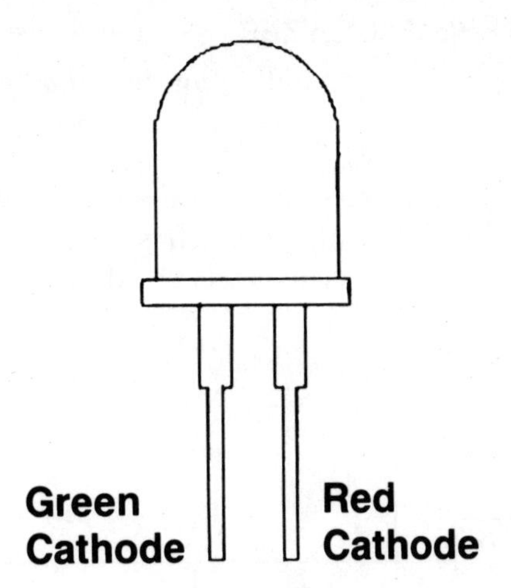

Fig. 5-13. Pin assignments for RCA SK2026.

Product Example: RCA SK2026

Package Configuration: Bi-color red/green T-1 3/4 L-007

Forward Current: 35 mA/Green; 70 mA/red

Bandwidth: 560/green; 660/red

WAVE SHAPE ANALYZER

Aside from their use as logic indicators (ON or OFF), LEDs are able to duplicate the pixel representation from a low-resolution graphics display. In this application, each LED is equivalent to a single graphics screen picture element or *pixel*. An example of this use is presented with the Wave Shape Analyzer which is a 10 × 10 LED matrix based on a similar 5 × 7 LED matrix circuit that was described in the August 1979 issue of *Popular Electronics*.

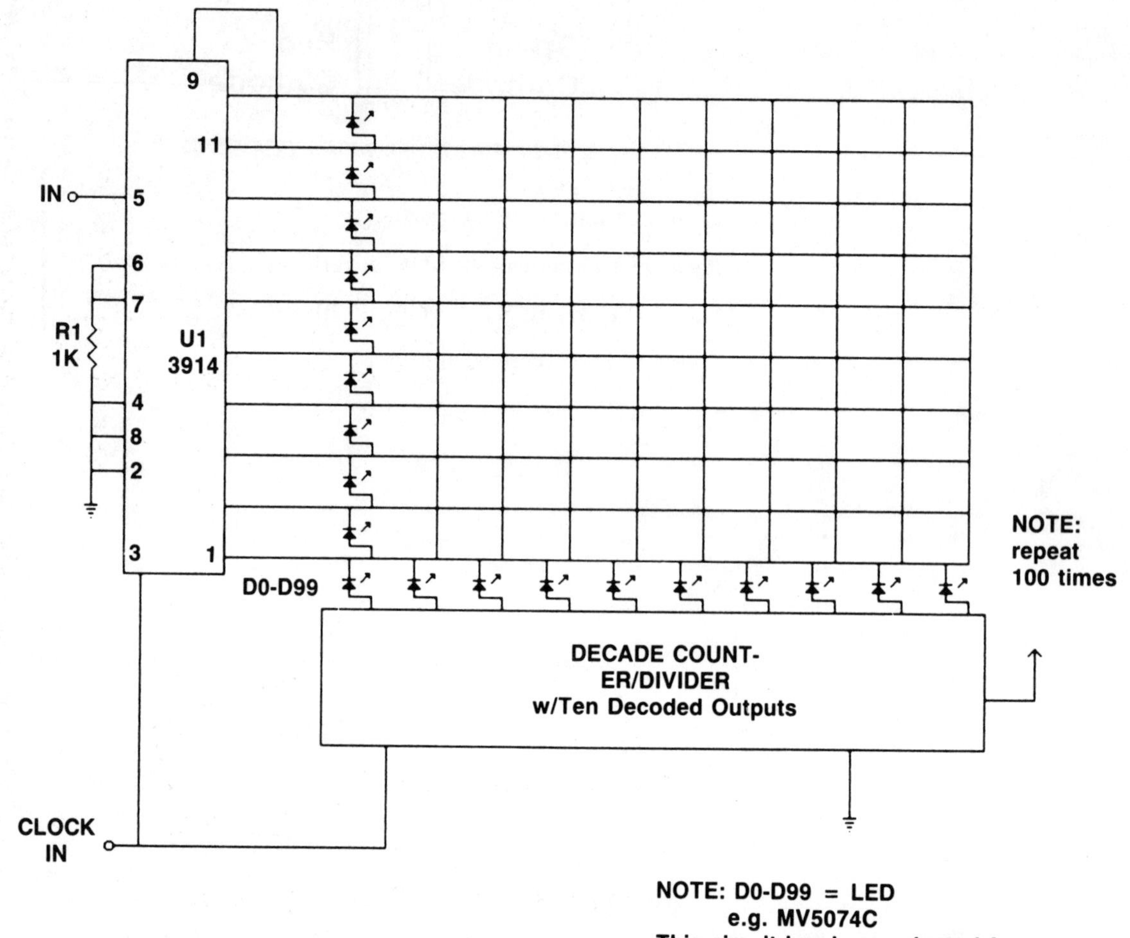

Fig. 5-14. Schematic diagram for Wave Shape Analyzer.

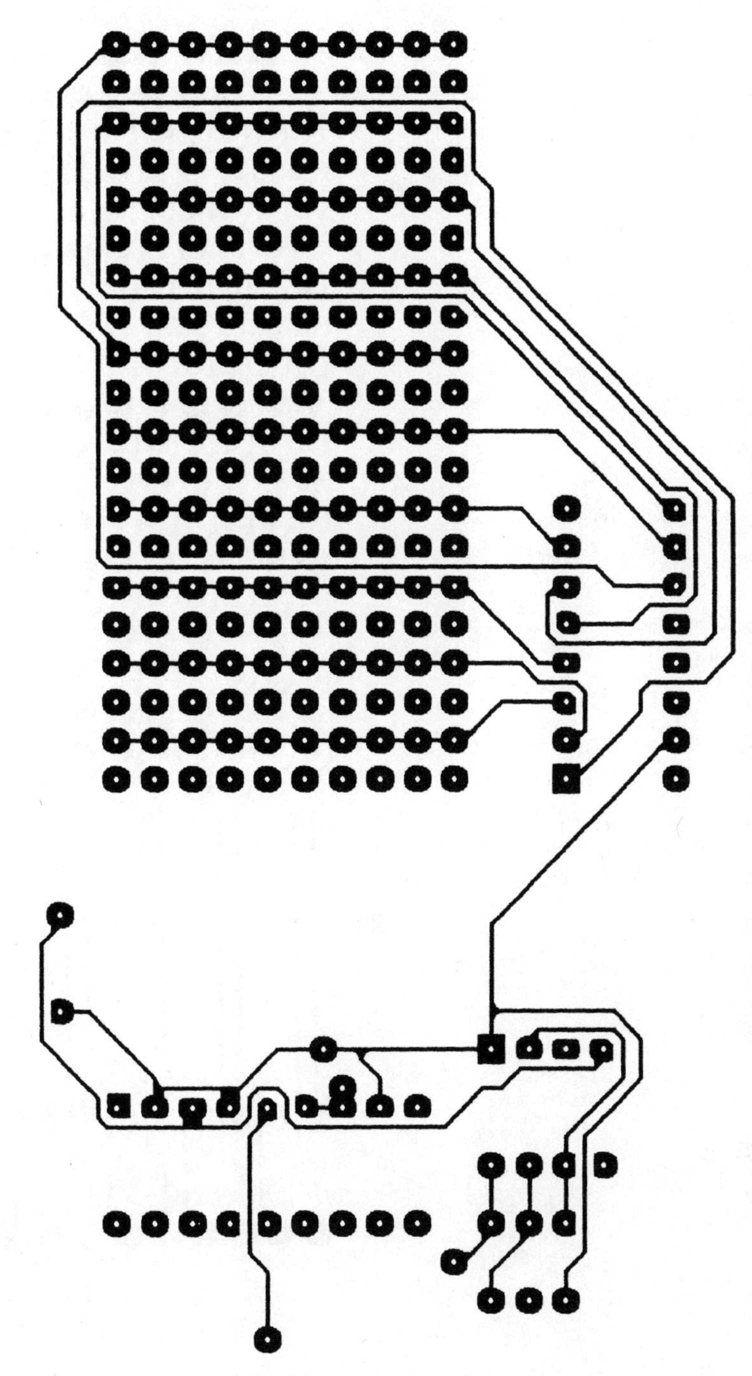

Fig. 5-15. Solder side for the Wave Shape Analyzer template shown at 2X size.

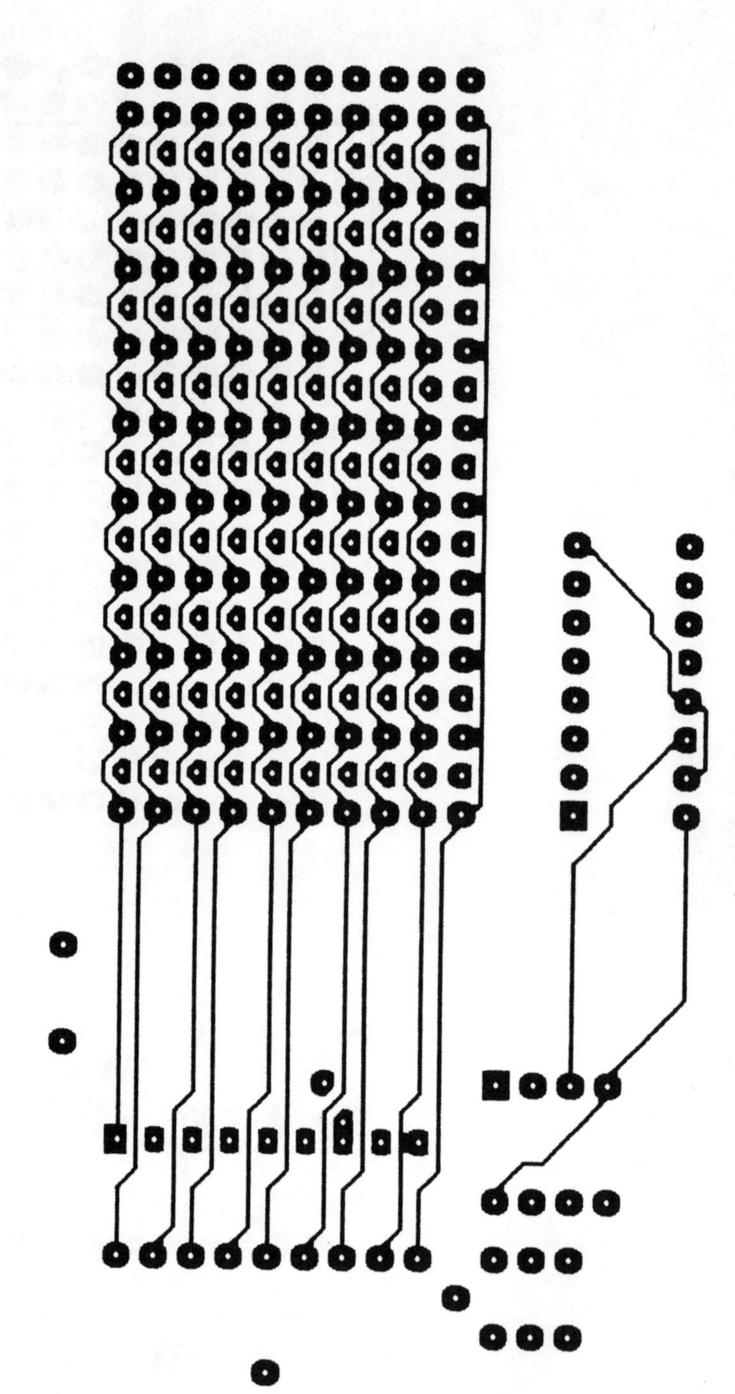

Fig. 5-16. Component side for the Wave Shape Analyzer template.

Construction Notes

Only three ICs are needed for generating a graphics representation of a waveshape input (see Fig. 5-14). Of these, the clock circuit sequentially pulses the outputs of the decade counter, while the dot driver IC (U1 3914) selectively triggers a specific row of LEDs. This triggering is based on the shape of the input waveform on pin 5 of the 3914.

The Wave Shape Analyzer is a continuously running graphics display. A clock-based potentiometer can be used for slowing or increasing the sweep of the display. In this configuration, virtually any form of wave input can be rudimentarily visualized.

During the assembly of the Wave Shape Analyzer, be sure to observe the polarity of the LEDs. Failure to adhere to this precaution will result in an error-plagued project. Of course, soldering the 200 leads for the matrix LEDs will be an extremely taxing procedure resulting in a nightmare of twisting and overlapping cables. One solution to avoiding this spaghetti wiring is to translate the Wave Shape Analyzer's schematic to a compact PCB design. Figures 5-15, 5-16, and 5-17 show two sides of a PCB template and a suggested parts layout, respectively. As illustrated in these templates, T-1 LEDs will fit in the provided pads. Using the larger T-1 3/4 LEDs would create a congested PCB and an inability to accommodate all 100 of the LEDs.

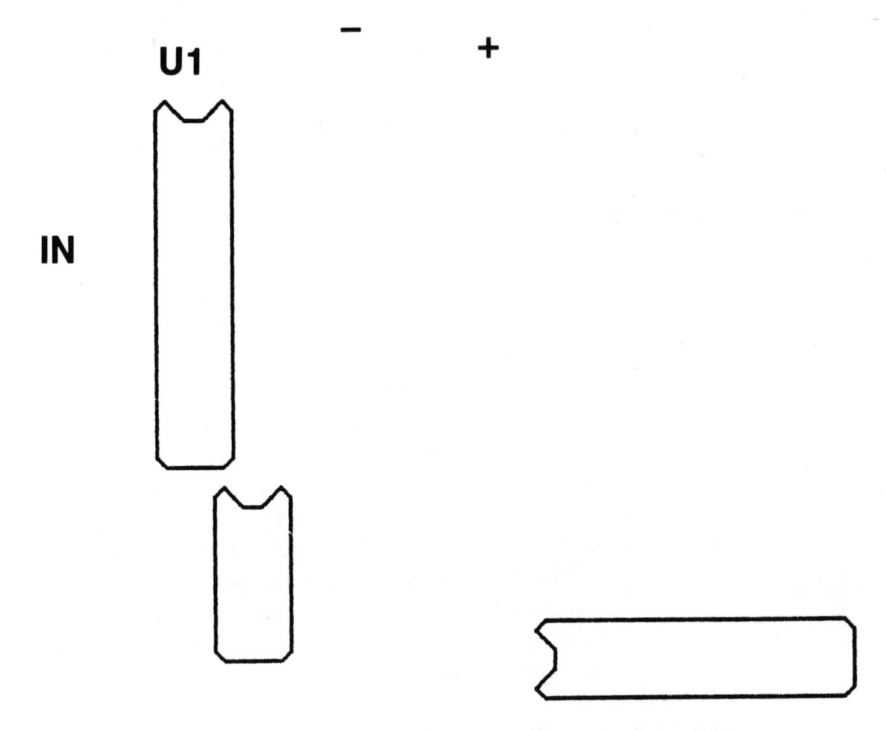

Fig. 5-17. Parts layout for the Wave Shape Analyzer PCB.

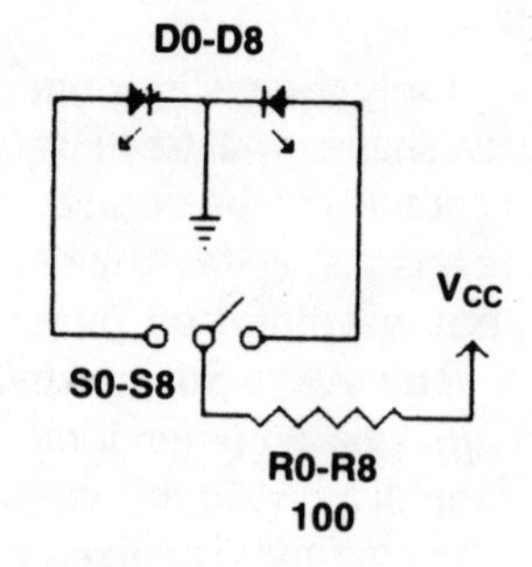

Fig. 5-18. Schematic diagram for Tic-Tac-Toe.

TIC-TAC-TOE

Three-terminal LEDs, like R9-56, usually house two different colored LEDs (e.g., red and green) inside the same plastic carrier. A white diffused lens makes the radiation of either the red or green photons equally brilliant from virtually any viewing angle. While the simultaneous lighting of both of these colored LEDs renders the multi-colored display pointless, a high-tech version of an old game can be designed by selectively emitting one of the colors singly.

Construction Notes

Tic-Tac-Toe is a simple circuit consisting of nine SPDT switches and nine corresponding three-terminal LEDs (see Fig. 5-18). In order to limit the forward current to the forward-biased LEDs, a 220-ohm resistor should be applied to the center lug of each SPDT switch.

During the operation of Tic-Tac-Toe, one player selects red and the other is assigned green. The red player moves first by flipping the switch to the red LED's anode. This action illuminates the red LED inside the three-terminal LED housing. The second player then lights the desired green LED by flipping the corresponding SPDT. This player interaction continues until the game ends in either a win or a draw. An interesting point about the Tic-Tac-Toe circuit is that it doesn't require a power switch. If each SPDT is left in its center position after the game is finished, then no current will flow through the circuit. This action both preserves the battery's life as well as eliminates the need for yet another switch.

6

LED Light Bars

The simplest display form of multiple LEDs is found in the light bar. Also known as *dot displays*, *bar displays*, and *LED arrays*, LED light bars combine two or more singular LEDs inside the same housing. These primitive graphics displays lack the complexity and sophistication of multi-segment displays (see Chapter 7), but they are quite capable of converting analog data into a readable digital format.

A typical analog signal that can easily be displayed by an LED light bar is the audio strength level of an input volume (see Fig. 6-1). In this application, the light bar is being used as an audio VU meter (volume unit meter).

The circuit that is used for driving this light bar VU meter consists of ten separate LED driver circuits. In turn, each of these drivers is connected to a separate comparator whose noninverting input is controlled by a scaling resistor. The inverting input of each comparator is connected in parallel to an input voltage buffer.

When no signal is being sent to the input buffer, all of the LEDs are inactive. As the signal is applied to the buffer, the reference voltage of the first comparator drives its output into a high condition. This activates the corresponding LED driver which lights the connected LED. If the signal voltage continues to increase, each of the sequential LED drivers/comparators go high and light its connected LED.

The format of the LED lighting is determined by two factors: the threshold established by the *scaling resistors* and *window comparators*. Scaling resistors can be used for selecting either a linear or a logarithmic display of

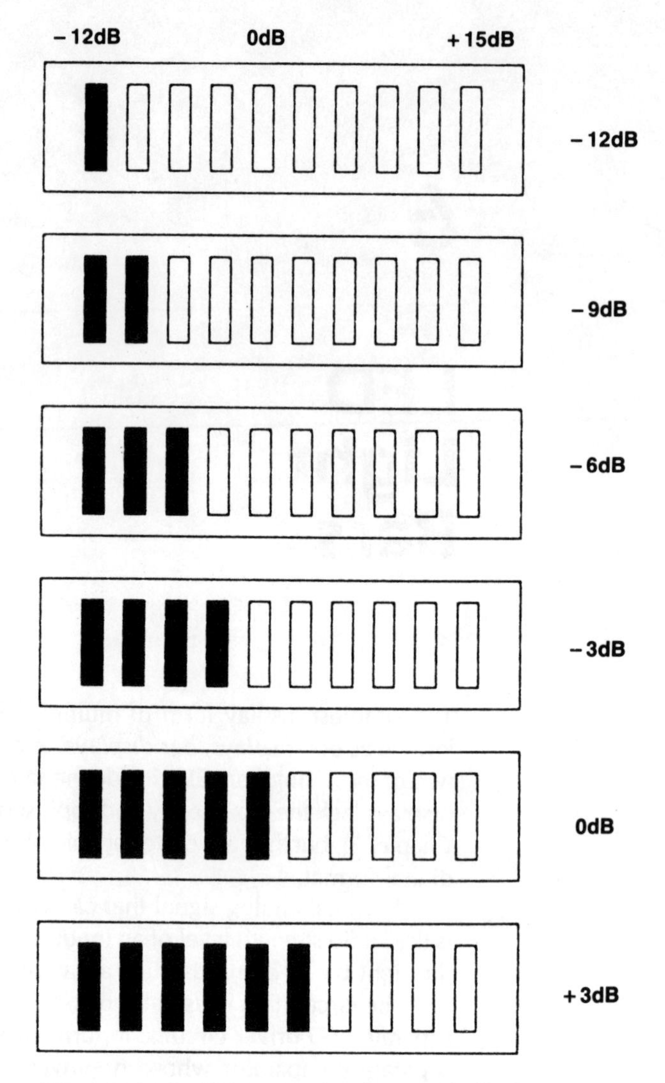

Fig. 6-1. Audio VU meter signal strength displays.

the input signal voltage. These displays are fixed by the resistance values of the scaling resistors.

The other method for controlling the LED lighting format is through window comparators. A window comparator is only able to light its connected LED when the input voltage lies between a predetermined high and low value; hence, the term "window" comparator. Therefore, when window comparators are used in an LED light bar, only one LED is ever lit at a single time. This action results in a moving dot display as opposed to the bar display obtained from using standard voltage comparators.

LIGHT BARS

Assembled in 2-, 4-, and 8-LED arrays, light bars are offered in either SIP (*s*ingle *i*n-line *p*ackage) or DIP carriers. In all cases, each of the light bar LEDs is controlled via a separate anode and cathode pin. Therefore, any combination of LEDs can be illuminated within the light bar.

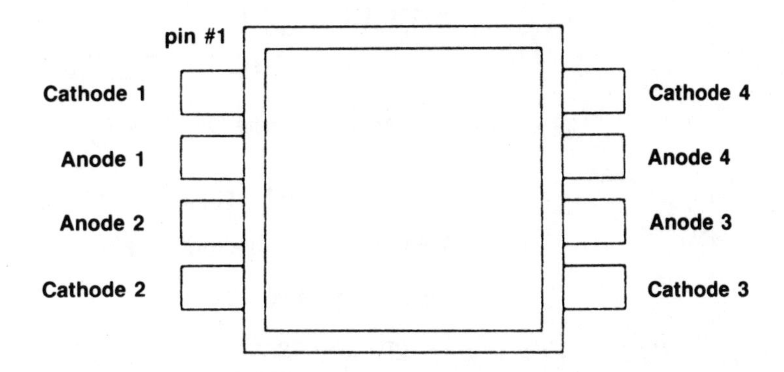

Fig. 6-2. Pin assignments for General Instrument HLMP-2655.

Product Example: General Instrument HLMP-2655

Package Configuration: 4-LED DIP

Forward Voltage: 2.6 V

Reverse Voltage: 6.0 V

Wavelength: 630 nm

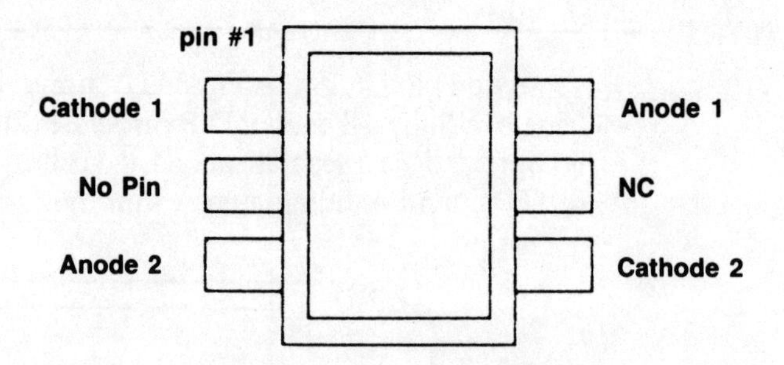

Fig. 6-3. Pin assignments for General Instrument MV53173.

Product Example: General Instrument MV53173

Package Configuration: 2-LED DIP

Forward Voltage: 2.5 V

Reverse Voltage: 5.0 V

Wavelength: 585 nm

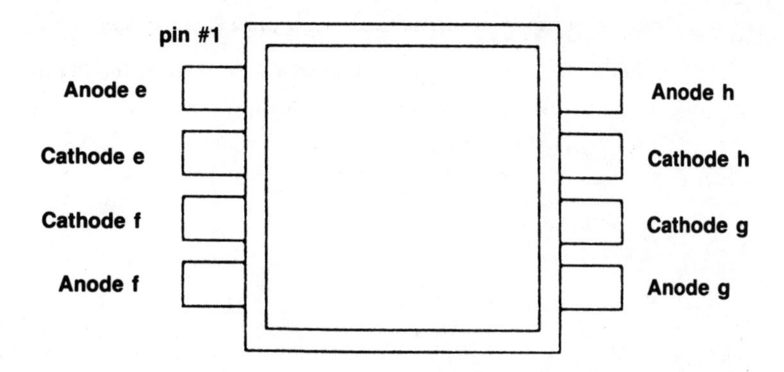

Fig. 6-4. Pin assignments for Hewlett-Packard HLMP-2965, RCA SK2117, and B1001R.

Product Example: Hewlett-Packard HLMP-2965

Package Configuration: Bi-color, 2-LED, 8-pin DIP

Forward Voltage: 2.6 V

Wavelength: Red 635 nm; green 565 nm

BAR GRAPH ARRAYS

These multi-element (10 or more LEDs) arrays are housed in DIPs with a linear arrangement designed for optimal viewing of signal information in either a bar or a dot display. An end-stacking feature makes larger element displays an easy interfacing task.

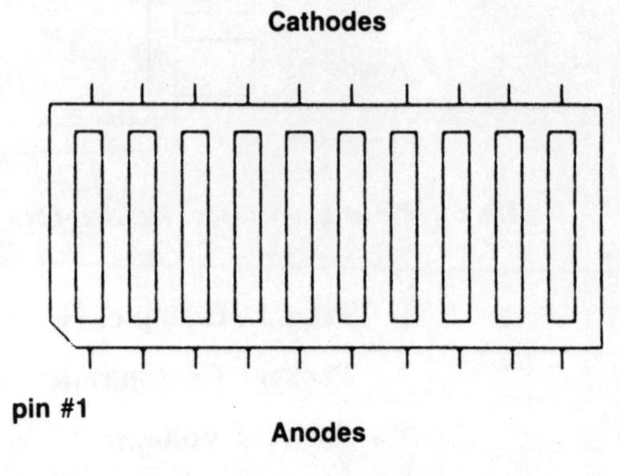

Fig. 6-5. Pin assignments for General Instrument MV57164.

Product Example: General Instrument MV57164, RCA SK2117, and B1001R

Package Configuration: 10-element 20-pin DIP

Forward Voltage: 3.0 V, 3.0 V, and 1.6 V, respectively

Reverse Voltage: 6.0 V, 6.0 V, and 5.0 V, respectively

Wavelength: 630 nm, 585 nm, and 655 nm, respectively

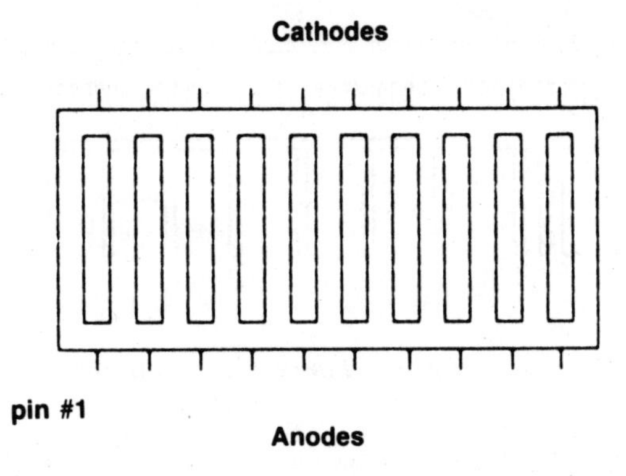

Fig. 6-6. Pin assignments for Hewlett-Packard HDSP-4840.

Product Example: Hewlett-Packard HDSP-4840

Package Configuration: 10-element 20-pin DIP

Forward Voltage: 2.5 V

Reverse Voltage: 3.0 V

Wavelength: 583 nm

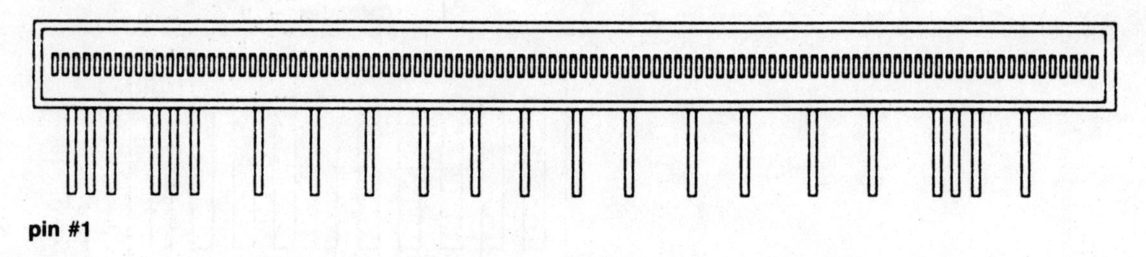

Fig. 6-7. Pin configuration for Hewlett-Packard HDSP-8820.

Product Example: Hewlett-Packard HDSP-8820

Package Configuration: 101-element 37-pin SIP

Internal Logic Arrangement: 10 common cathode blocks of 10 LED segments

Forward Voltage: 2.1 V

Reverse Voltage: 3.0 V

Wavelength: 640 nm

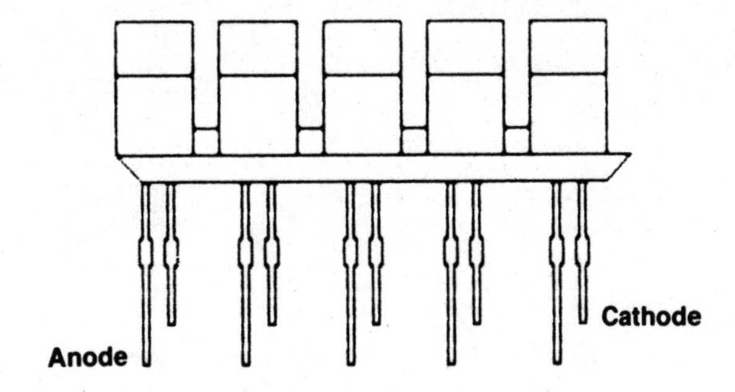

Fig. 6-8. Pin assignments for RCA SK2154.

Product Example: RCA SK2154

Package Configuration: 5-element 10-pin SIP

Forward Voltage: 1.9 V

Reverse Voltage: 3.0 V

Wavelength: 700 nm

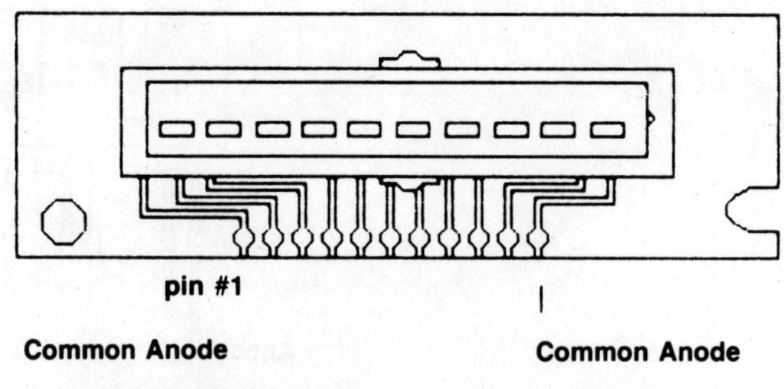

Fig. 6-9. Pin configuration for RCA SK2114.

Product Example: RCA SK2114

Package Configuration: 9-element 11-pin SIP

Internal Logic Arrangement: Two common anode pins

Forward Voltage: 2.05 V

Reverse Voltage: 3.0 V

Wavelength: 565 nm

AUDIO VU METER

By interfacing an LED bar/dot graphic display (see Fig. 6-10) to a 3916 display driver IC, a low-cost 10-element VU meter can be constructed. Basically, this IC can receive analog voltage signals and drive up to ten LEDs in a graphic display of this input. Either a bar or a moving dot display can be selected through one pin of the 3916. This results in a powerful VU meter that can be selected through one pin of the 3916. This results in a powerful VU meter that can handle any form of data display.

Construction Notes

Only one support component is needed for building the Audio VU Meter (see Figs. 6-11, 6-12, and 6-13). The addition of another resistor, a potentiometer, and a capacitor increases the response, LED brightness, and reduces circuit noise of the Audio VU Meter.

As presented, the Audio VU Meter is a bar display. If a dot display is desired, then wire pin 9 of the 3916 to ground. Furthermore, a 1.2 V internal reference voltage is required by the Audio VU Meter. Another resistor should be added between pin 8 and ground for increasing the limit for this reference voltage.

Fig. 6-10. A ten-element LED graphic display versus a two-element LED light bar. Notice the internal wiring of the two-element light bar (far right).

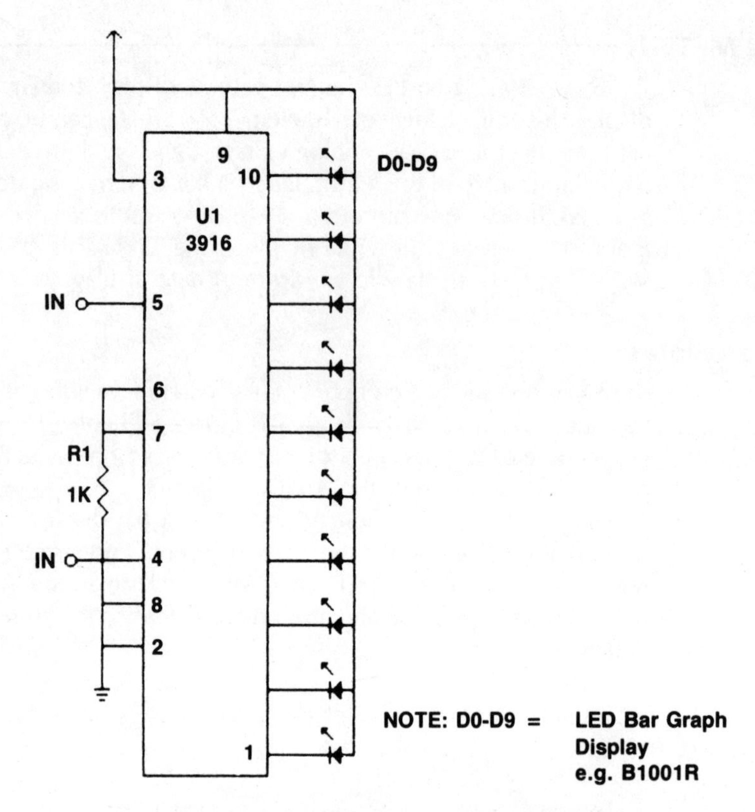

Fig. 6-11. Schematic diagram for Audio VU Meter.

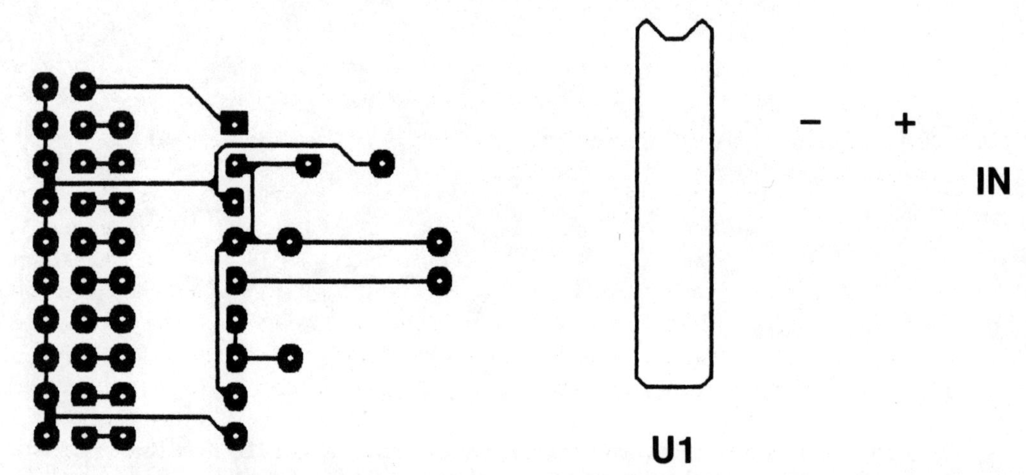

Fig. 6-12. Solder side for the Audio VU Meter template, shown at 2X size.

Fig. 6-13. Parts layout for the Audio VU Meter PCB.

7

Multi-Segment LED Displays

Representing alphanumeric characters through gallium arsenide technology can be accomplished with two general types of LED displays: segmented and dot matrix. Although both of these display types can be used for duplicating a complete range of alphabetic and numeric characters, they reach their final visual product along different paths.

A segmented LED display consists of several light-emitting segments arranged in a fixed planar pattern. Selectively stimulating one or more of these segments causes the display to form a single and specific character. Based on the target system's character-generation parameters, there are three segmented LED display configurations to choose from: 7-segment, 14-segment, and 16-segment (see Fig. 7-1). Each of these display types offers a greater degree of character resolution as a result of increasing the number of segments.

Dot-matrix LED displays, on the other hand, are typically available in only one configuration: a 5×7 grid or matrix (see Fig. 7-2). In this format, all of the characters that can be represented by any of the segmented displays (including the 16-segment display) can be formed with a single dot-matrix LED display. In order to achieve this degree of character resolution, the structure of the dot-matrix display differs significantly from the fixed planar pattern utilized by segmented displays. A dot-matrix display consists of 35 individual LEDs linked into common anode and common cathode rows and columns. One common industry representative of this format is the MAN27 5×7 dot-matrix display. In this example, each column is formed by seven

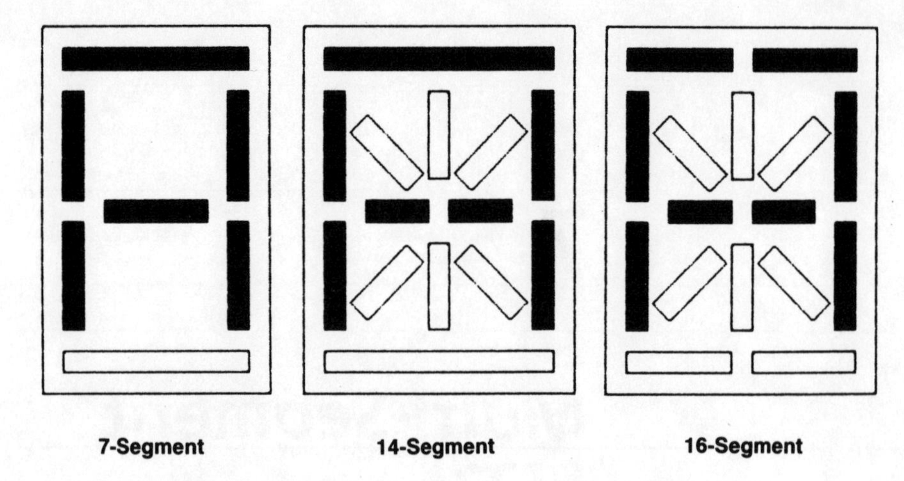

Fig. 7-1. Character resolution comparisons between 7-segment, 14-segment, and 16-segment LED displays.

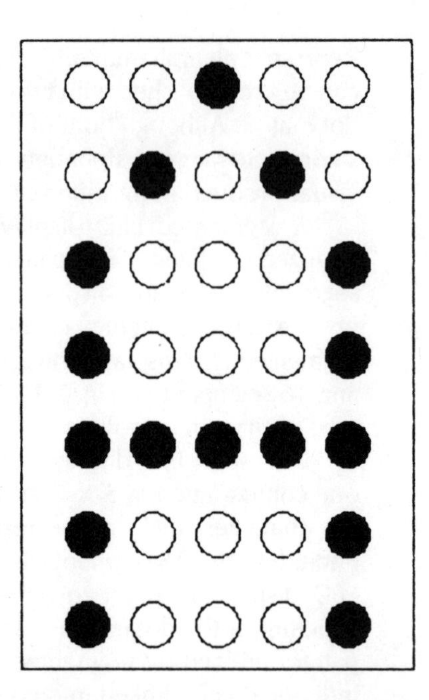

Fig. 7-2. Character resolution of a 5 × 7 dot-matrix LED display.

linked anodes, while a row is formed by five linked cathodes. A project later in this chapter suggests how to build a MAN27 display.

One important facet of using either segmented or dot-matrix displays is the need for a display driver circuit. This often complex circuitry handles the correct and timely lighting of the LED segments or dots that are needed for representing a desired character. Generally speaking, there are two practical methods for lighting multi-segment LED displays (in this context, this term applies equally to either segmented or dot-matrix displays): *pulsing* and *multiplexing*.

The simplest of these two drive methods is a pulsing circuit. Basically, the pulsing circuit is a low-cost means of strobing a display rapidly on and off. The visual result of this strobing is a steadily glowing display with a low power-consumption rate. When used in single digit or character displays, this method is adequate. In larger, multi-character displays, however, multiplexing is the only solution for reducing both the display driver circuit component count as well as reducing power consumption.

Multiplexing a multi-digit, segmented, LED display requires three basic logic circuits: a counter, a decoder, and a clock. In this configuration, the pulses from the clock circuit control the input of BCD (*b*inary-coded *d*ecimal) counter data into a decoder. This decoder then selectively drives a single digit and its associated segment decoder circuitry. The result from this action is that only individual segments or digits are actually illuminated at any given time. Fortunately, by multiplexing the steady stream of clock pulses, this singular data is strobed so rapidly that the group of digits appear to glow steadily to the slower human eye.

Another slightly different multiplexing method is used for driving a dot-matrix display. In this case, a ROM (*r*ead-*o*nly *m*emory) circuit along with a shift register are used for rapidly strobing the display's rows and columns into producing a character. This top-to-bottom and left-to-right progression generates the visual appearance of a glowing, fully formed character. There are two strobing techniques for multiplexing a dot-matrix display: horizontal and vertical strobing. These two techniques differ only in the orientation and placement of the decoding ROM and the shift register. In practical applications, horizontal strobing is the simpler of these two multiplexing techniques and is usually limited to displays with fewer than five characters. Conversely, vertical strobing involves a more sophisticated level of circuit design but drives a greater number of individual characters or digits.

SEGMENTED DISPLAYS

Available in numerous sizes, colors, and resolutions, segmented displays are limited in their diversity of character representations based on the resolution of their segmentation. The higher resolution segmented displays offer a limited ability to display specialty characters, while the lower resolution 7-segment displays are restricted to displaying ten numbers and nine distinct alphabetic letters.

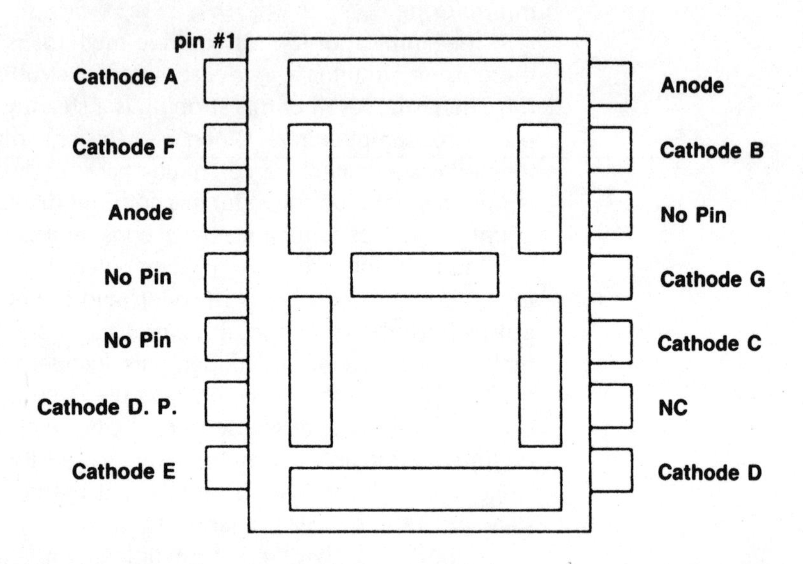

Fig. 7-3. Pin assignments for General Instrument 5082-7650 and MAN1A.

Product Example: General Instrument 5082-7650 and General Instrument MAN1A

Package Configuration: 14-pin DIP

Individual Character Size: 0.43-inch character and 0.27-inch character, respectively

Segment Number: 7-segment

Forward Voltage: 2.5 V and 4.0 V, respectively

Reverse Voltage: 3.0 V and 10.0 V, respectively

Wavelength: 630 nm

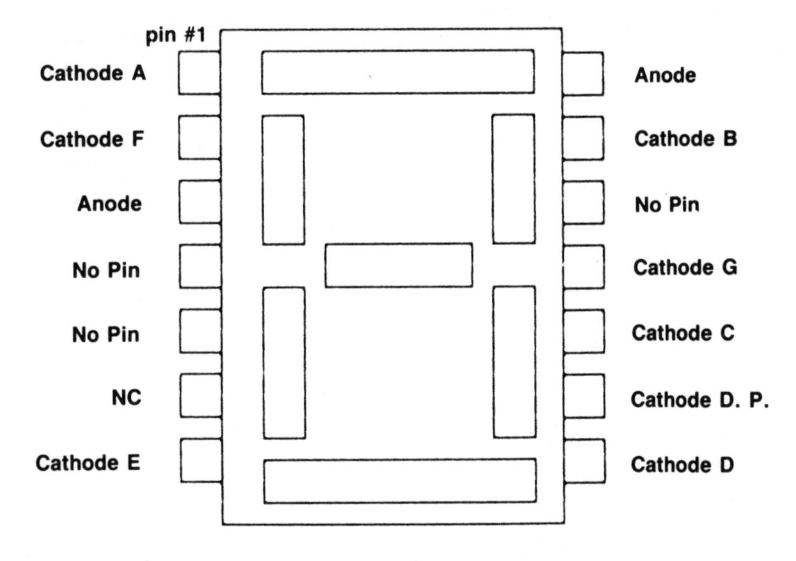

Fig. 7-4. Pin assignments for General Instrument MAN3610A.

Product Example: General Instrument MAN3610A

Package Configuration: 14-pin DIP

Individual Character Size: 0.30-inch character

Segment Number: 7-segment

Forward Voltage: 2.5 V

Reverse Voltage: 6.0 V

Wavelength: 630 nm

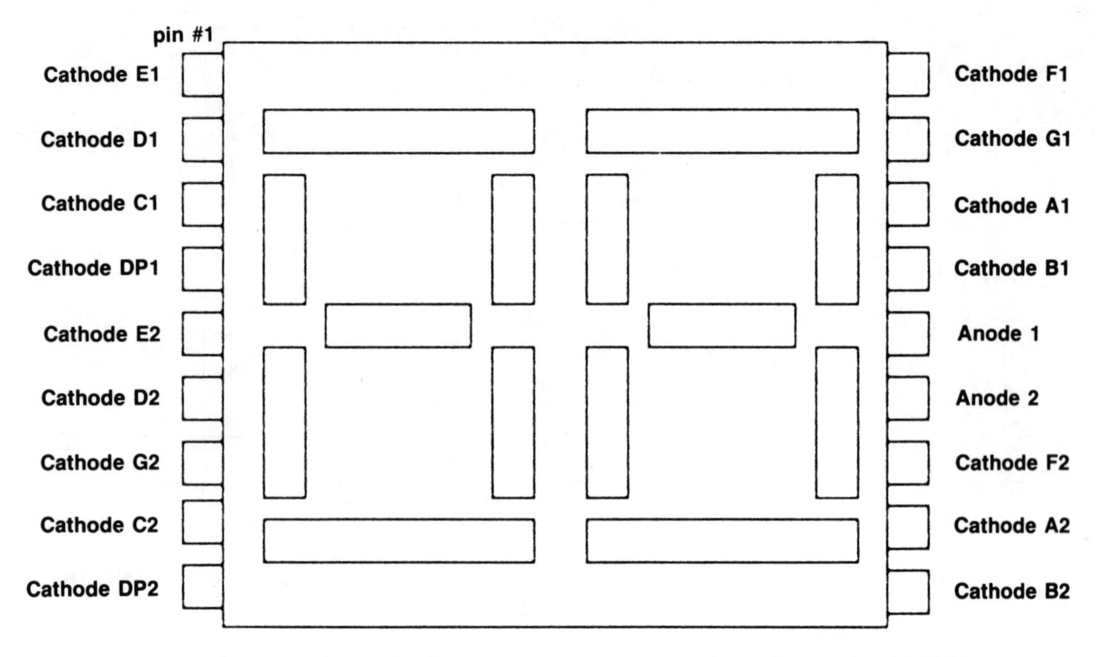

Fig. 7-5. Pin assignments for General Instrument MAN6110 and Hewlett-Packard HDSP-5721.

Product Example: General Instrument MAN6110 and Hewlett-Packard HDSP-5721

Package Configuration: 18-pin DIP

Individual Character Size: Two 0.56-inch characters

Segment Number: 7-segment

Forward Voltage: 2.2 V and 2.5 V, respectively

Reverse Voltage: 6.0 V and 3.0 V, respectively

Wavelength: 635 nm and 583 nm, respectively

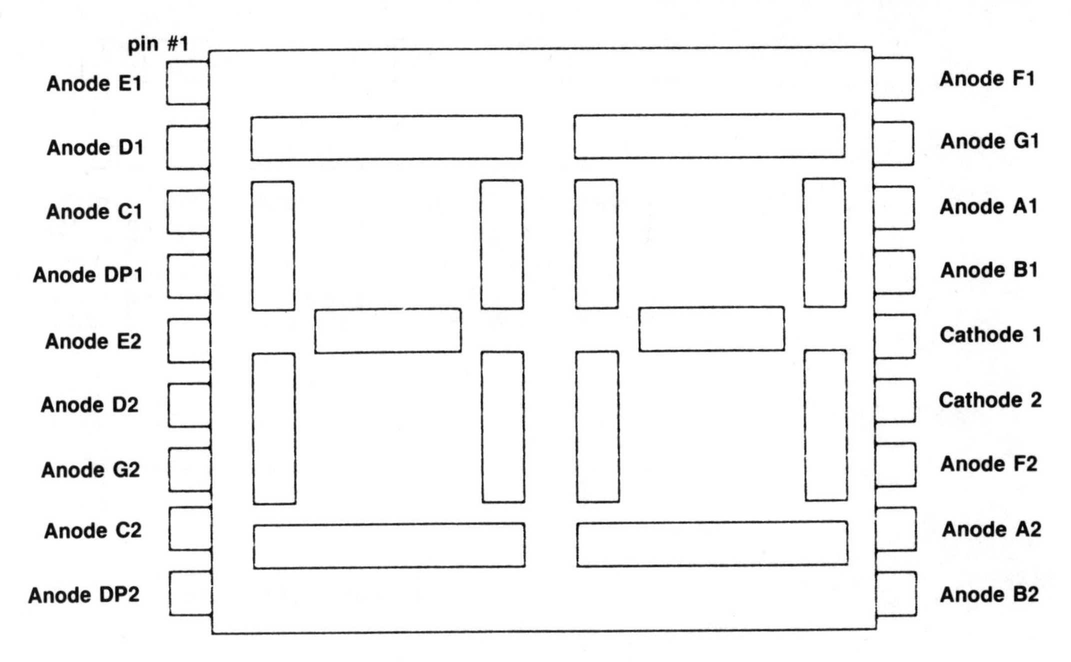

Fig. 7-6. Pin assignments for General Instrument MAN6440.

Product Example: General Instrument MAN6440

Package Configuration: 18-pin DIP

Individual Character Size: Two 0.560-inch characters

Segment Number: 7-segment

Forward Voltage: 2.2 V

Reverse Voltage: 6.0 V

Wavelength: 562 nm

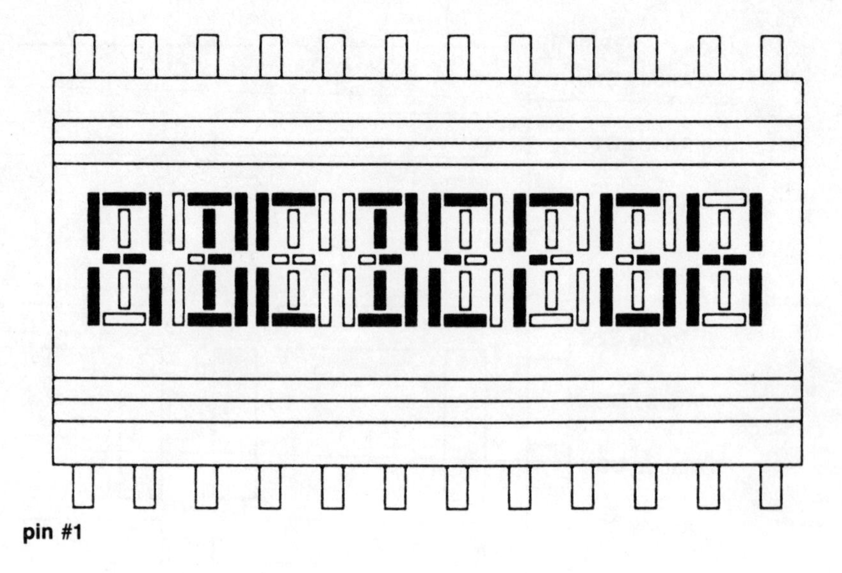

Fig. 7-7. Pin configuration for General Instrument MAN2815.

Product Example: General Instrument MAN2815

Package Configuration: 24-pin DIP

Individual Character Size: Eight 0.135-inch characters

Segment Number: 14-segment

Forward Voltage: 2.0 V

Reverse Voltage: 5.0 V

Wavelength: 660 nm

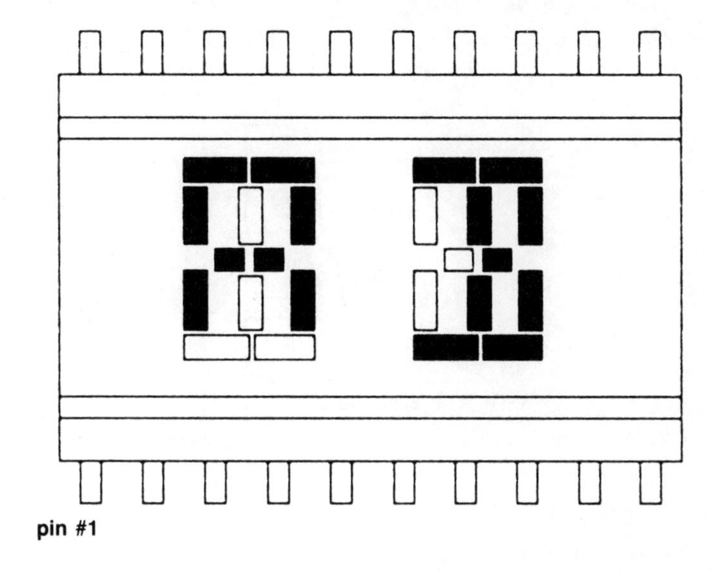

Fig. 7-8. Pin configuration for General Instrument MMA58420.

Product Example: General Instrument MMA58420

Package Configuration: 20-pin DIP

Individual Character Size: Two 0.5-inch characters

Segment Number: 16-segment

Forward Voltage: 2.5 V

Reverse Voltage: 5.0 V

Wavelength: 585 nm

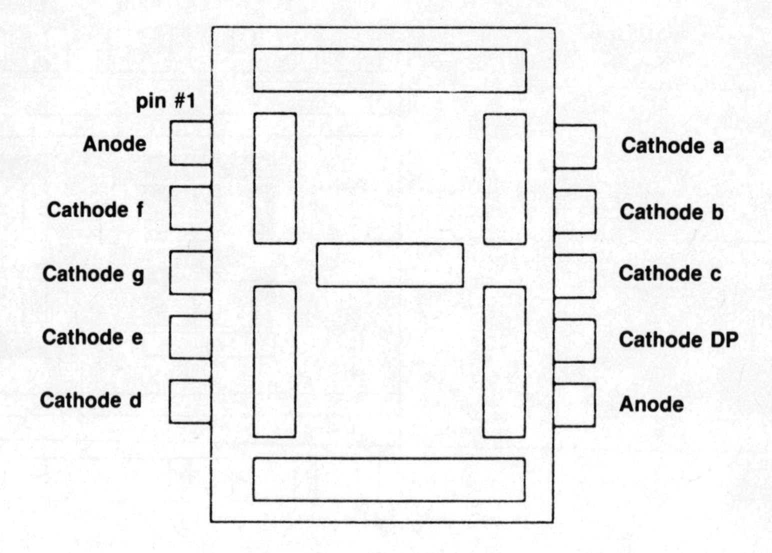

Fig. 7-9. Pin assignments for Hewlett-Packard HDSP-7801.

Product Example: Hewlett-Packard HDSP-7801

Package Configuration: 10-pin DIP

Individual Character Size: 0.3-inch character

Segment Number: 7-segment

Forward Voltage: 2.5 V

Reverse Voltage: 3.0 V

Wavelength: 566 nm

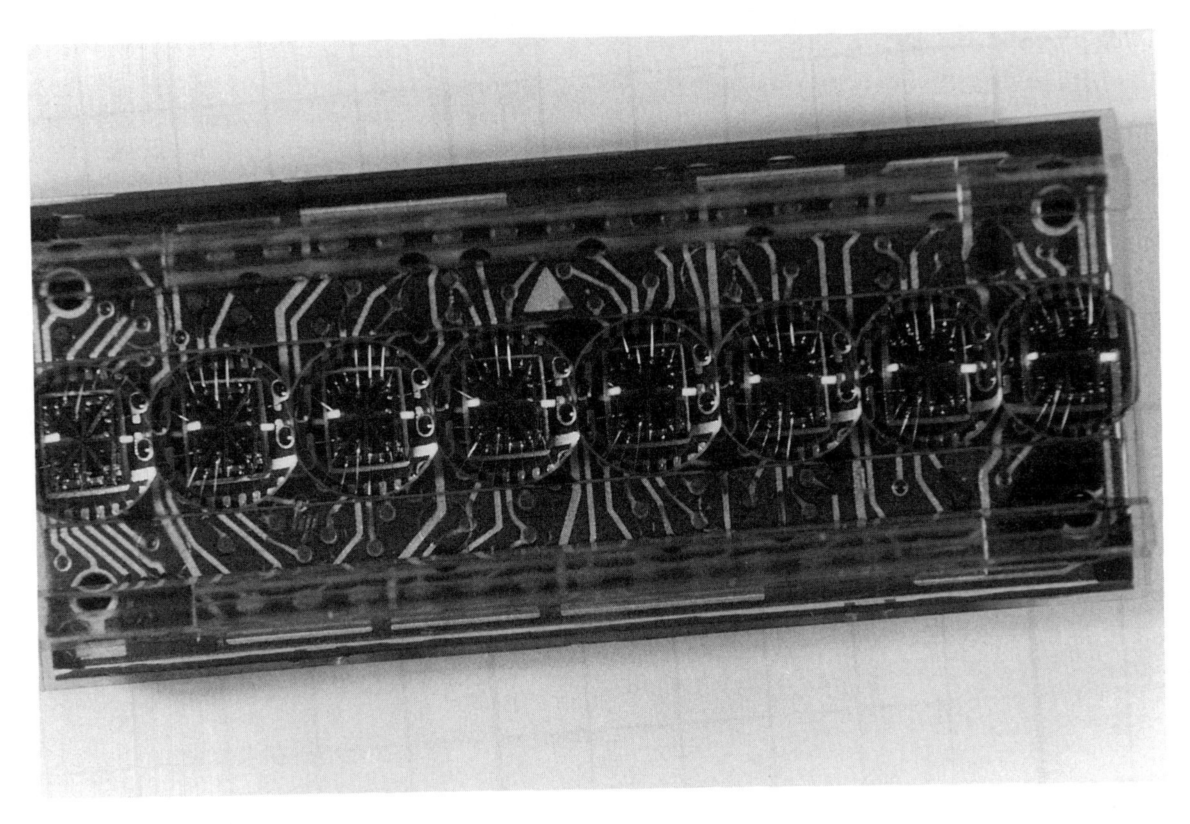

Fig. 7-10. Pin configuration for Hewlett-Packard HDSP-6508.

Product Example: Hewlett-Packard HDSP-6508

Package Configuration: 26-pin DIP

Individual Character Size: Eight 0.15-inch characters

Segment Number: 16-segment

Forward Voltage: 1.9 V

Reverse Voltage: 5.0 V

Wavelength: 655 nm

DOT-MATRIX DISPLAYS

Like segmented displays, dot-matrix displays come in various colors and sizes. Unlike their segmented counterparts, however, dot-matrix displays are manufactured in a fixed 5 × 7 matrix field of 35 individual LEDs. Furthermore, in this configuration, the dot-matrix display is able to generate all ten digits, a complete upper and lower case alphabet, and complex graphics and special characters.

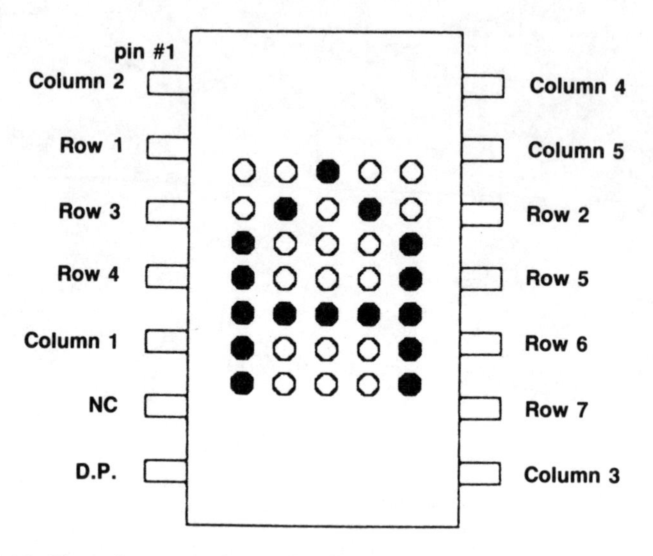

Fig. 7-11. Pin assignments for General Instrument MAN2A and MAN27.

Product Example: General Instrument MAN2A and MAN27

Package Configuration: 14-pin DIP

Individual Character Size: 0.32-inch character

Forward Voltage: 2.0 V and 1.8 V, respectively

Reverse Voltage: 5.0 V

Wavelength: 660 nm

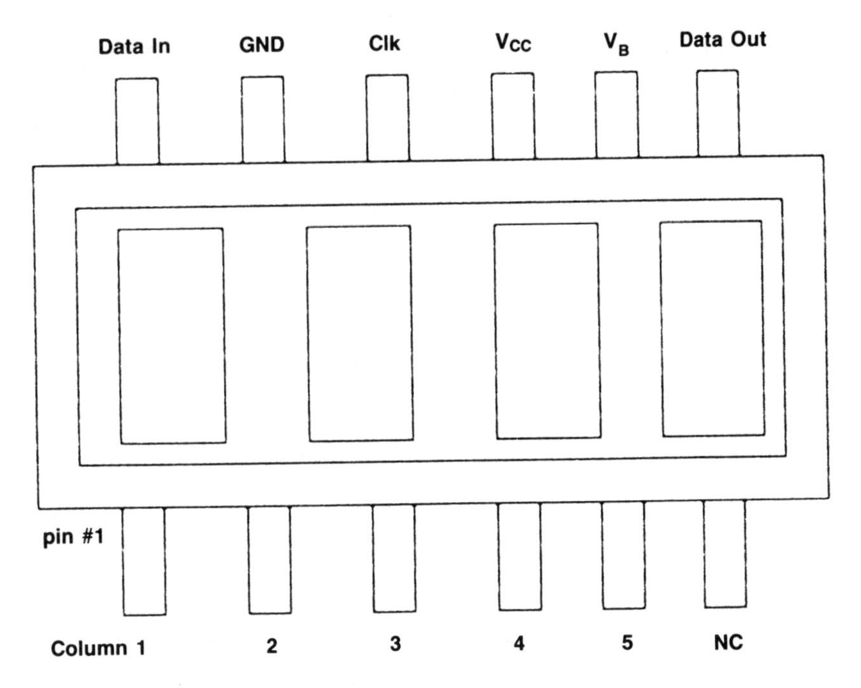

Fig. 7-12. Pin assignments for Hewlett-Packard HDSP-2000.

Product Example: Hewlett-Packard HDSP-2000

Package Configuration: 12-pin DIP

Individual Character Size: Four 0.15-inch characters

Forward Voltage: 2.0 V

Reverse Voltage: 5.0 V

Wavelength: 655 nm

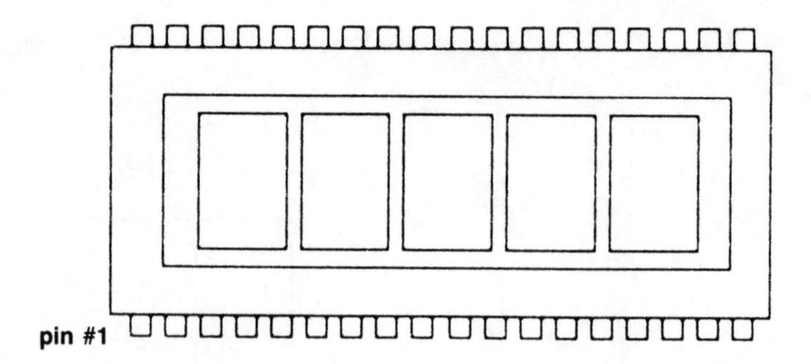

Fig. 7-13. Pin configuration for Hewlett-Packard HDSP-7102.

Product Example: Hewlett-Packard HDSP-7102

Package Configuration: 36-pin DIP

Individual Character Size: Five 0.6-inch characters

Forward Voltage: 2.0 V

Reverse Voltage: 4.0 V

Wavelength: 655 nm

VOLTMETER

Intersil manufactures a series of display ICs that are capable of driving a segmented LED display. The first of these display ICs is the ICL7107. This is an A/D (analog / digital) converter chip that can directly drive a 3½-digit 7-segment LED display.

Construction Notes

This Voltmeter uses a handful of support components for converting a voltage input into a digital display (see Fig. 7-14). All of the other active display logic circuitry, decoders, drivers, and clock are self-contained within the ICL7107. In addition to the support components, three common anode 7-segment displays, along with a common anode overflow display are used for displaying the voltage input.

During operation of the Voltmeter, two separate supply voltages are required by the ICL7107 (U1). These +5.0 V and −5.0 V voltages are applied to pins 1 and 26, respectively. A voltage variation to a maximum of +6.0 V and −9.0 V is tolerable.

COUNTER

Another valuable, self-contained Intersil display driver IC is the ICM7226B. This IC is a multi-function universal counter and 7-segment LED display driver. Based on the status of three control pins—function input, range input, and control input—the presentation of the data on the LED display can be configured for 16 different functions and/or ranges.

Construction Notes

The control functions for the Counter can be either hardwired (see Fig. 7-15) or manipulated through a series of rotary switches. As presented in Fig. 7-15, the Counter is a unit counter (pin 4), operating with a range of 1 sec /100 cycles (pin 21), and with an external oscillator selectively enabled or disabled.

Only digital data can be accepted by the two input ports (pin 2 and pin 40). The rate of input frequency is fixed by the oscillator on pins 35 and 36 at 10 MHz. In order to input other forms of data, buffer and amplifier stages need to be added to these signals prior to interfacing them to the ICM7226B's (U1) input ports. Once the input data has been properly configured, a battery of eight 7-segment LED displays are used for displaying these data.

MICROPROCESSOR DISPLAY INTERFACE

One final Intersil IC, the ICM7243A, can drive either 14- or 16-segment LED displays through a microprocessor interface. On board the ICM7243A are a 64-character ASCII (american standard code for information interchange) decoder, an 8 × 6 memory, character and segment drivers, and a multiplexing circuit.

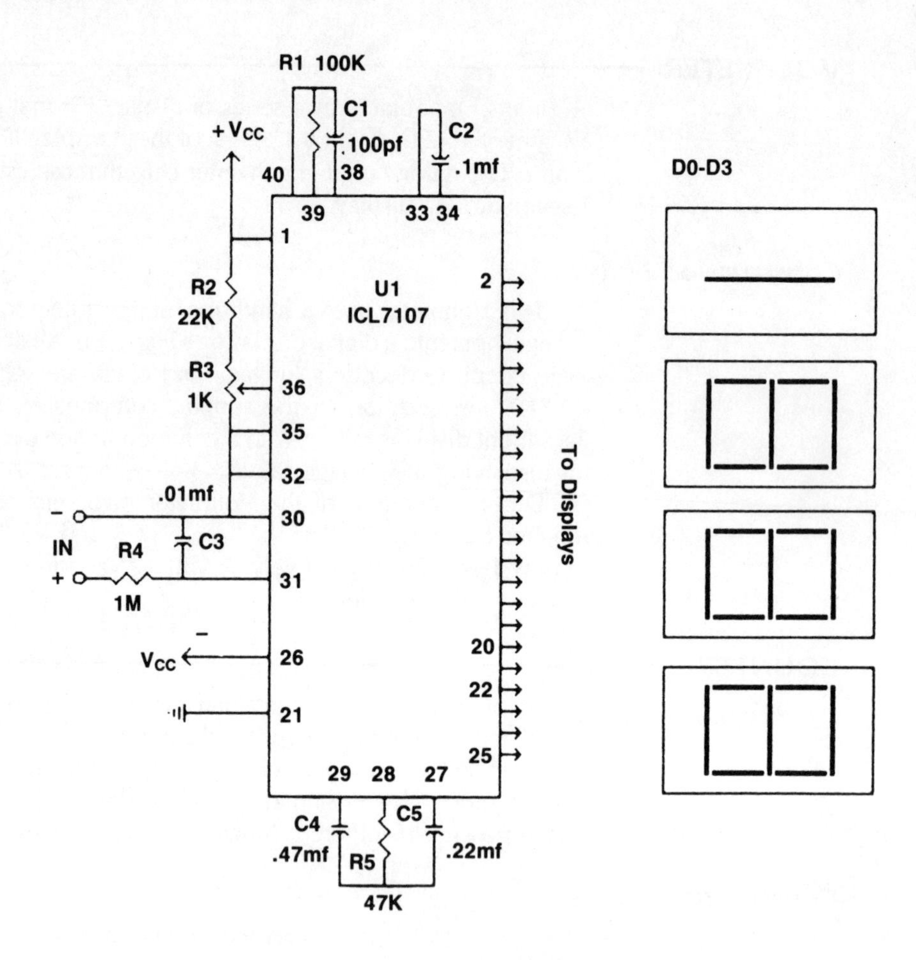

Fig. 7-14. Schematic diagram for Voltmeter.

Construction Notes

Remarkably, the Microprocessor Display Interface requires no additional support components for operation (see Fig. 7-16). Only a six-bit ASCII data input is required for displaying on either a 14- or 16-segment, 8-digit LED display.

An internal mode latch selects the method of displaying the input data on the LED display. Pin 31 of the ICM7243A (U1) determines whether a serial access mode (a high state) or a random access mode (a low state) has been

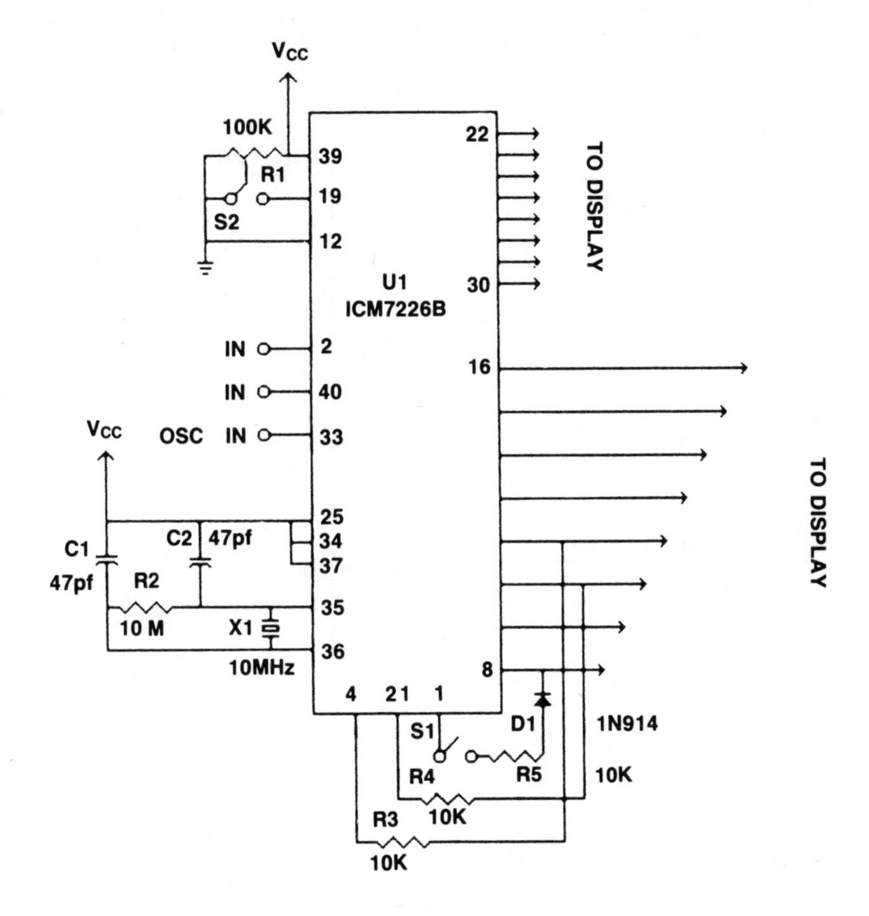

Fig. 7-15. Schematic diagram for Counter.

externally selected. In the serial access mode, the first data entry is displayed in the left-most character position in a multi-digit display. Following this placement, each additional data entry is placed in the next higher position to the right of the previous digit.

Contrary to the serial access mode, the random access mode relies on the microprocessor for determining the placement for each data entry. The address pins (pins 28, 29, and 30) are used by the microprocessor for selecting this placement position.

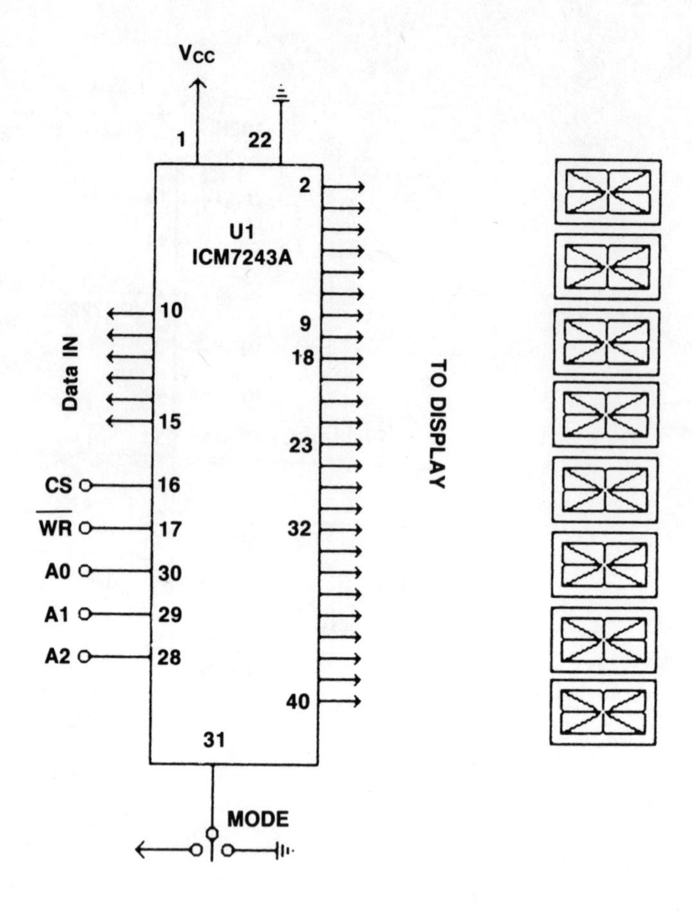

Fig. 7-16. Schematic diagram for Microprocessor Display Interface.

LED 5 × 7 TERMINAL

Working with dot-matrix displays presents two problems to the experimenter. First, the cost of each display is three times higher than a comparable segmented display. Second, the size of the dot-matrix display can be prohibitively small when placed in large format systems (e.g., scrolling LED displays). Therefore, an alternative is to design a simple dot-matrix display based on the popular configurations found in commercial displays.

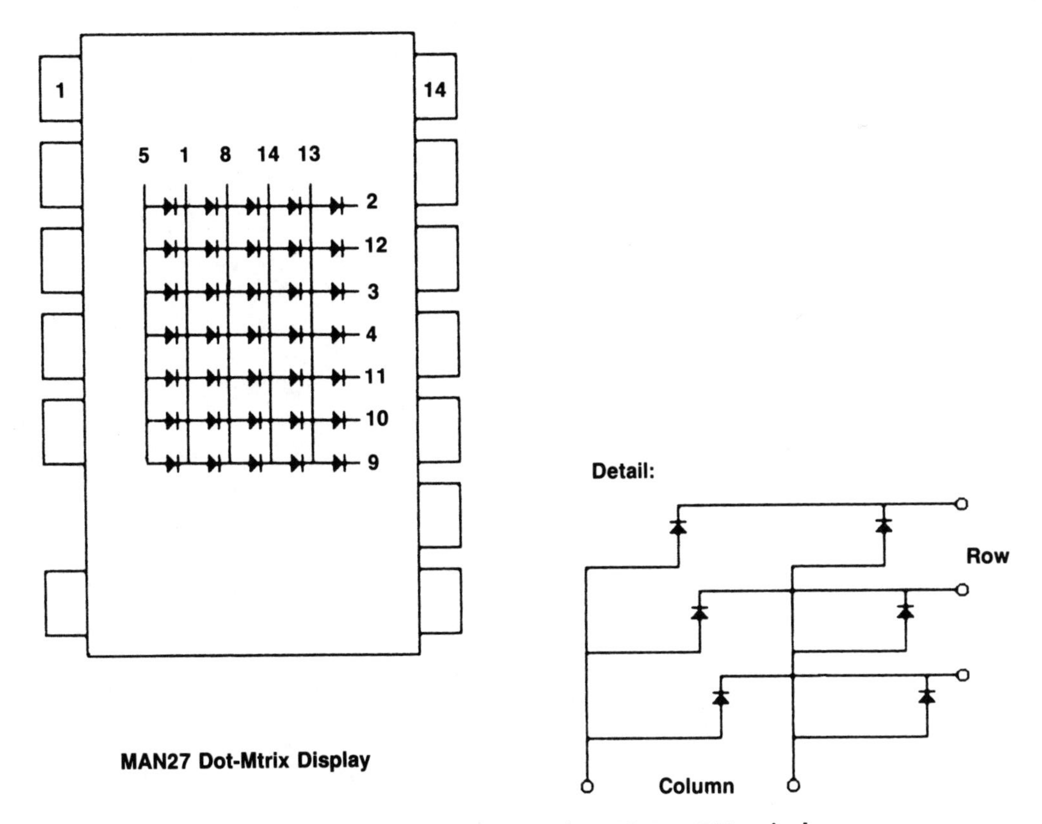

Fig. 7-17. Schematic diagram for LED 5 × 7 Terminal.

Construction Notes

The LED 5 × 7 Terminal is a simple 14-pin dot-matrix display based on the MAN27 pin configuration (see Fig. 7-17). In order to reduce the size of the completed display, 35 T-1 size discrete LEDs should be used in the construction of the LED 5 × 7 Terminal (see Fig. 7-18). Otherwise, standard T-1¾ size discrete LEDs can be used in the 5 × 7 matrix.

Assembly time for the LED 5 × 7 Terminal can be reduced by using a PCB template (see Figs. 7-19 and 7-20). This technique also permits the construction of several dot-matrix displays for building larger, more complex displays.

Both horizontal strobing and vertical strobing multiplexing can be used with the LED 5 × 7 Terminal. A 74LS164 serial shift register (along with an inverter/buffer IC) makes an effective means of interfacing the LED 5 × 7 Terminal to a microprocessor. In this format, a serial signal could be sent from the controlling microprocessor and displayed on the LED 5 × 7 Terminal.

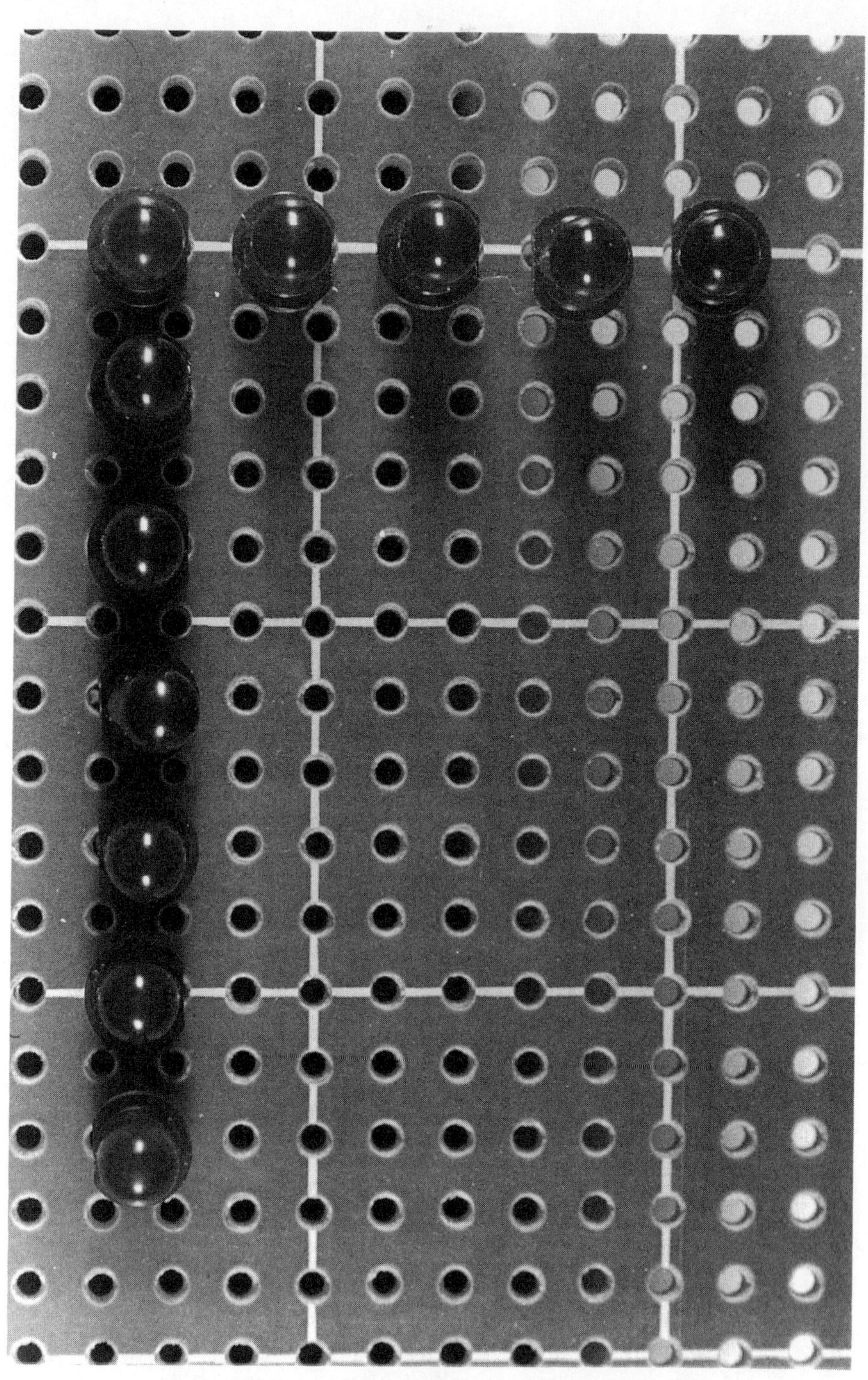

Fig. 7-18. Using T-1 LEDs for the construction of the LED 5 × 7 Terminal provides adequate spacing around each LED.

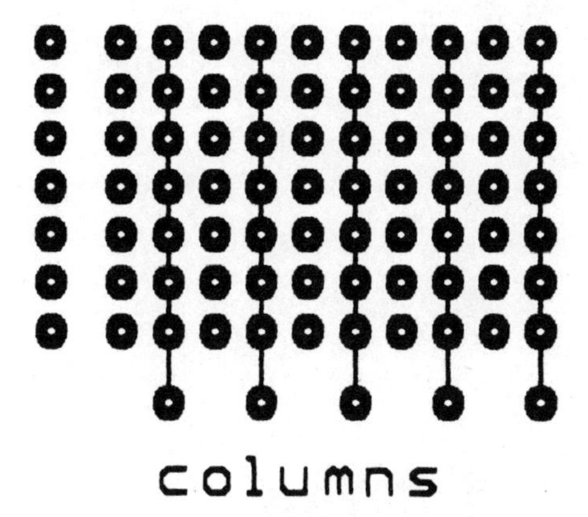

Fig. 7-19. Solder side for the LED 5 × 7 Terminal shown at 2X size.

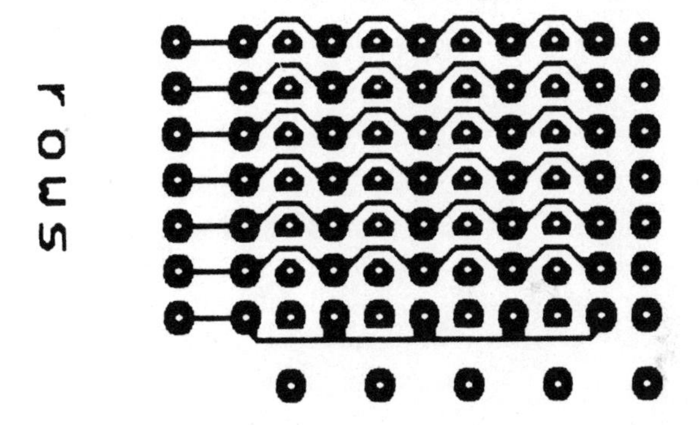

Fig. 7-20. Component side for the LED 5 × 7 Terminal.

8

GaAs Laser Diodes

The light generated when current is injected into a gallium arsenide substrate excites a photon radiation popularly known as *laser* (*l*ight amplification by *s*timulated emission of *r*adiation) *pulsations*. This photon emission differs significantly from the forms of light radiated by other members of the gallium arsenide family.

In general, light production from lasers is typified by a monochromatic, coherent, collimated wavelength. This results in a finely focused beam of light that is unaffected by the *inverse square law* of light diffusion over a given distance factor. Likewise, the member photons of this collimated light source are generated through the stimulated emission of photon-to-atom controlled strikes. An atom population inversion state must exist, however, before the density of these stimulated emissions exceeds the more natural spontaneous emission photon generation condition.

Inverting this atom population is achieved by the electronic pumping of high-level currents into the GaAs junction. Most GaAs laser diodes used are forward biased p/n diodes with two metal contacts linked to a power supply. This link is used for forcing current into the p/n junction through current injection. By using current injection in a GaAs laser diode, the light output from the diode is proportional to the current density at the junction. Therefore, an injection laser has a minimum current density for initiating the lasing process.

Lasing from the GaAs laser diode is achieved through the cleaved end at the junction's edge. This lasing process is in direct contrast to that found in more conventional gas and solid-state lasers that rely on mirrors for generating the required amount of reflectivity needed for lasing. The mirrorless GaAs diode junction generates a reflectivity of roughly 36 percent. The generated lasing beam is then emitted from both ends of the junction. Painting one end of the junction serves to increase the reflectivity of the open junction end.

One performance-negating by-product of GaAs laser diode photon radiation is a rise in temperature. This increase in temperature serves to decrease the output of the laser. There are three areas where the output of the GaAs laser is retarded with an increase in temperature: power output, wavelength output, and beam divergence. The relationship between a rise in temperature and each of these output characteristics is expressed as:

Temperature increase =

Power decrease
Wavelength increase (i.e. longer wavelengths)
Beam Divergence increase

One method of reducing the temperature production of this laser is through current pulsing. Pulsing a GaAs laser enables greater current to be applied to the junction. In turn, this pulsing reduces the diode's operating temperature, increases the power output, shortens the wavelength emissions, and decreases the beam's divergence.

Two calculations should be used during GaAs laser diode pulsing: the power average and the duty cycle. The power average represents the even distribution of power applied during a given PRT (*p*ulse *r*epetition *t*ime) or the time cycle needed for one pulse. This is expressed as:

$$P = E/PRT$$

where,

P = power average in watts
E = energy of each pulse in joules
PRT = frequency of each pulse in seconds

The duty cycle of a GaAs laser diode is equal to the ratio of the pulse width between the half-power points to the PRT. Therefore, the duty cycle can be found by:

$$DC = PW/PRT$$

where,

DC = duty cycle as a percentage
PW = pulse width in milliseconds
PRT = pulse frequency in milliseconds

LASER DIODE

A low threshold current is needed in GaAs substrate laser diodes for generating a high power output. These devices are continuous wave, high-temperature lasers that are ideal for pulsing applications.

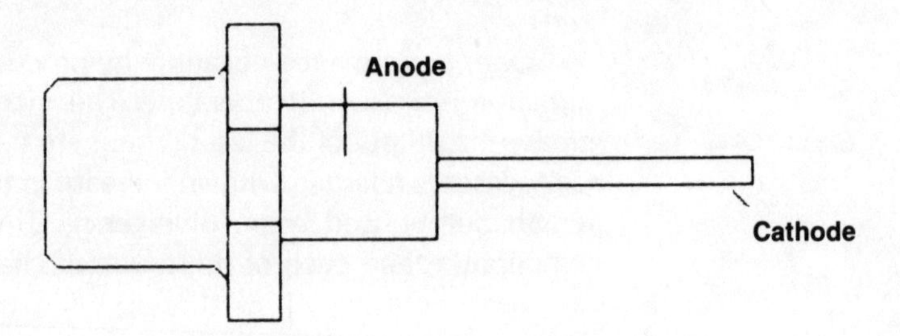

Fig. 8-1. Pin assignments for LCW-10.

Product Example: M/A-COM LCW-10

Package Configuration: LDL-9

Radiant Power Output: 14 mW

Reverse Voltage: 2.0 V

Forward Voltage: 2.0 V

Peak Wavelength: 830 nm

Threshold Current: 90 mA

Maximum Operating Temperature: 70°C

SIMPLE PULSED LASER

Caution: This project emits a concentrated beam of IR light that can be dangerous to the human eye. Exercise extreme care when using this circuit. *Never* look directly into the laser beam.

Construction Notes

A simple GaAs laser diode pulsing circuit can be created from a single transistor (see Fig. 8-2). Several support resistors and a capacitor are used for controlling the pulses in this project.

Before attempting to use the Simple Pulsed Laser, a careful measurement of the true pulse rate should be performed. A pulse rate under 100 ns should be used with this project. This rate will limit the power output to less than one watt. If a greater power output is required, then a more complex continuous wave (CW) circuit will be necessary.

COMPLEX CW INJECTION LASER

Caution: This project emits a concentrated beam of IR light that can be dangerous to the human eye. Exercise extreme care when using this circuit. *Never* look directly into the laser beam.

Construction Notes

A continuous wave GaAs laser does not require current limiting pulses that will, in turn, reduce the power output (see Fig. 8-3). Instead, a steady beam of light is emitted from the laser diode.

Due to the elimination of pulsing in this design, special attention must be paid to observing the proper values for the forward voltage, reverse voltage, and maximum operating temperature of the laser diode in this project. Exceeding any of these values will result in damage to the laser diode, as well as potential optic damage to an unprotected human eye. A suggested meth-

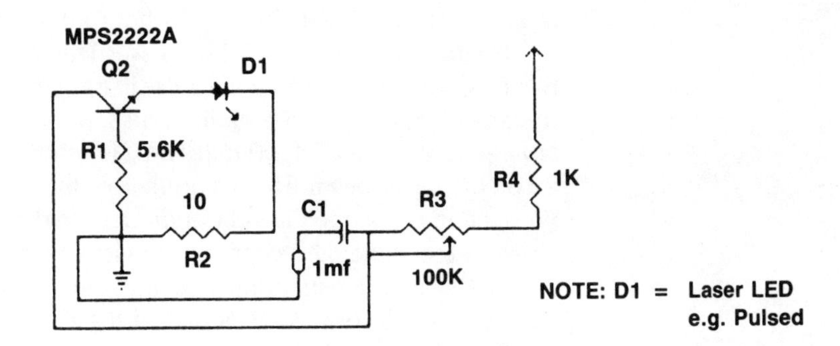

Fig. 8-2. Schematic diagram for Simple Pulsed Laser.

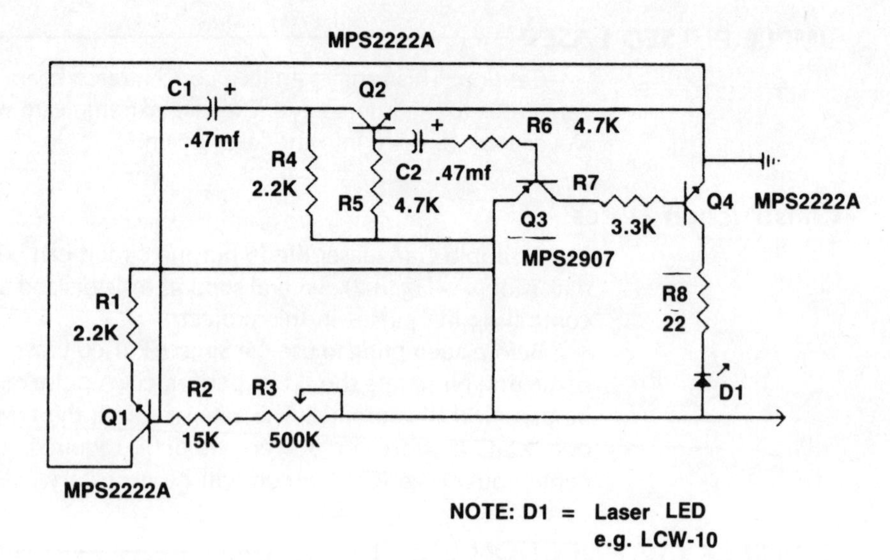

Fig. 8-3. Schematic diagram for Complex CW Injection Laser.

od for dealing with the limitations imposed by these three diode specifications is through the use of a filtered, regulated, voltage-controlled power supply. Furthermore, constantly monitoring the status of the laser diode through random voltage and temperature samples at the diode's cathode and anode will help in ensuring both the safety and the successful operation of this project.

HEATH ET-4200 LASER TRAINER KIT

Numerous potential hazards surround the construction of a functional homemade laser. Central among these problems is the assembly of a high-voltage power supply that is necessary for driving the laser. A low-cost kit from Heath Company eliminates this worry and reduces the construction of a functional laser to a simple 2½ hour circuit-board soldering exercise.

Heath's laser kit, the ET-4200 Laser Trainer, differs significantly from the two former laser projects cited earlier in this chapter. While the two previously described projects used a gallium arsenide laser diode for generating their laser emission, the ET-4200 relies on a low-power helium-neon gas laser tube for producing its beam. The net result from these two varying methods of beam production is that the GaAs laser diodes yield an IR light beam, whereas the Laser Trainer generates a light with a visible red wavelength of 632 nm. Other than this wavelength difference, the remainder of the Et-4200's design and construction closely parallels that of the GaAs projects.

Even though it is classified as a low yield, DHHS (Department of Health and Human Services) Class II, "educational" laser, Heath's Laser Trainer sports some impressive performance specifications. Activation of the helium-neon laser tube is initiated through a 4200-volt voltage tripler. Following this initial firing, the voltage tripler is deactivated and a −1400-volt voltage doubler is used as the primary power supply for the laser tube.

Coupled to the firing laser tube is a series of amplifying and modulating transistors. These circuits are capable of varying the intensity of the laser tube's output with a 10 percent modulation based on the input of either an external microphone (with a bandwidth of 300 Hz to 40 kHz) or an external 1-volt peak-to-peak signal source. The effect from this modulation is the transmission of messages through the ET-4200's laser beam to a receiving unit (e.g., Heath's ETA-4200 Laser Receiver). Other Laser Trainer specifications include:

- 0.49 mm beam diameter
- 1.64 mrad beam divergence
- random polarization
- 0.4 mW to 0.9 mW power output

The Heath ET-4200 Laser Trainer can be obtained in either a factory-wired form (ETW-4200) or built from a complete parts kit. While the assembly of a laser could plague the builder with power supply, beam alignment, and laser tube wiring nightmares, the ET-4200 is totally devoid of these preconceived construction pitfalls. The instruction and support provided in the detailed Heathkit manual transforms a potentially difficult kit construction into a simple step-by-step assembly process (see Fig. 8-4).

Thorough documentation has long been a strong selling point of Heath Company construction kits. In the case of the ET-4200, a 34-page assembly manual and a 22-page Illustration Booklet patiently guide the builder through all phases of construction, testing, and operation.

Construction begins with the soldering of 37 support components onto the PCB. Conveniently, Heath has supplied these components on a taped strip with each diode and resistor provided in the correct sequence for quick assembly (see Fig. 8-5). Following the correct installation of these parts, the remaining transistors, diodes, and capacitors are attached in their proper sequence. In short order, the Laser Trainer's PCB assembly is virtually complete.

Before the PCB is set aside, however, several large power capacitors and a power transistor must be attached to the PCB's component side. Similarly, the power transformer is soldered directly onto the circuit board (see Fig. 8-6). With the component side of the PCB now complete, the board is flipped over. Carefully, the helium-neon laser tube is mounted inside two holding clips and two wire connections are soldered onto its mounting lugs. Once this chore has been completed, the finished circuit board is set aside (see Fig. 8-7).

The work on the Laser Trainer now shifts to the assembly of the all-metal chassis. Like the PCB, the assembly of the chassis is a simple step-by-step

Fig. 8-4. Heath's documentation for the Heath ET-4200 Laser Trainer kit.

procedure. Following the attachment of several labels, the initial step involves the physical mounting of several jacks, a fuse holder, and a power switch. Each of these parts is supplied with separate mounting hardware. After each component has been securely fastened to the chassis, the completed PCB is wired to the various chassis-housed parts. Contrary to many other Heath kits, only nine wires must be prepared and soldered between the PCB and the chassis. Oddly enough, the greatest challenge during this final assembly of the chassis is the insertion of a strain relief around the power line cord. A strong pair of pliers and a healthy dose of persistence will finally seat the strain relief inside chassis hole AG. The completed chassis and PCB subassemblies are secured together with four screws as the final construction step.

Two adjustments must be made to the completed Laser Trainer. First, the laser beam is adjusted for optimal modulation. This procedure involves the operation of the Laser Trainer with its protective top cover removed (see Fig.

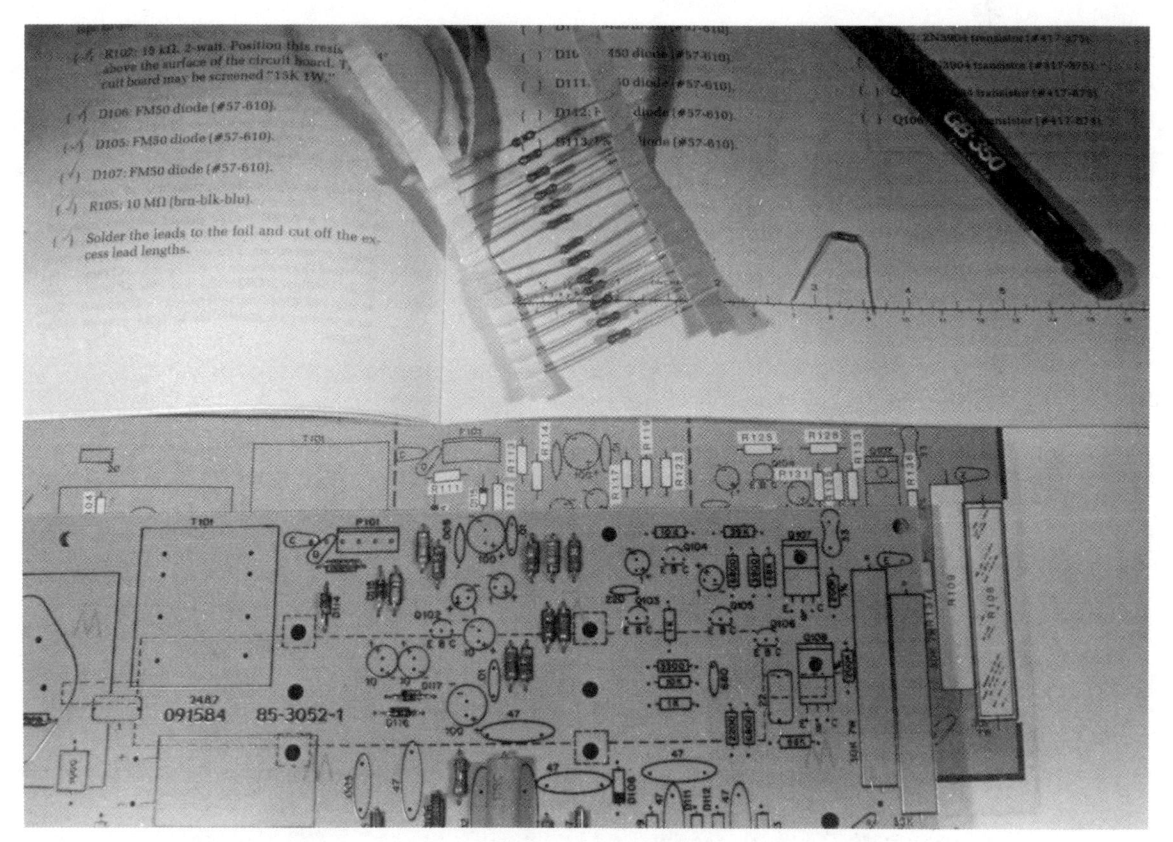

Fig. 8-5. Sequenced part tape strips make the soldering of these support components an error-free operation.

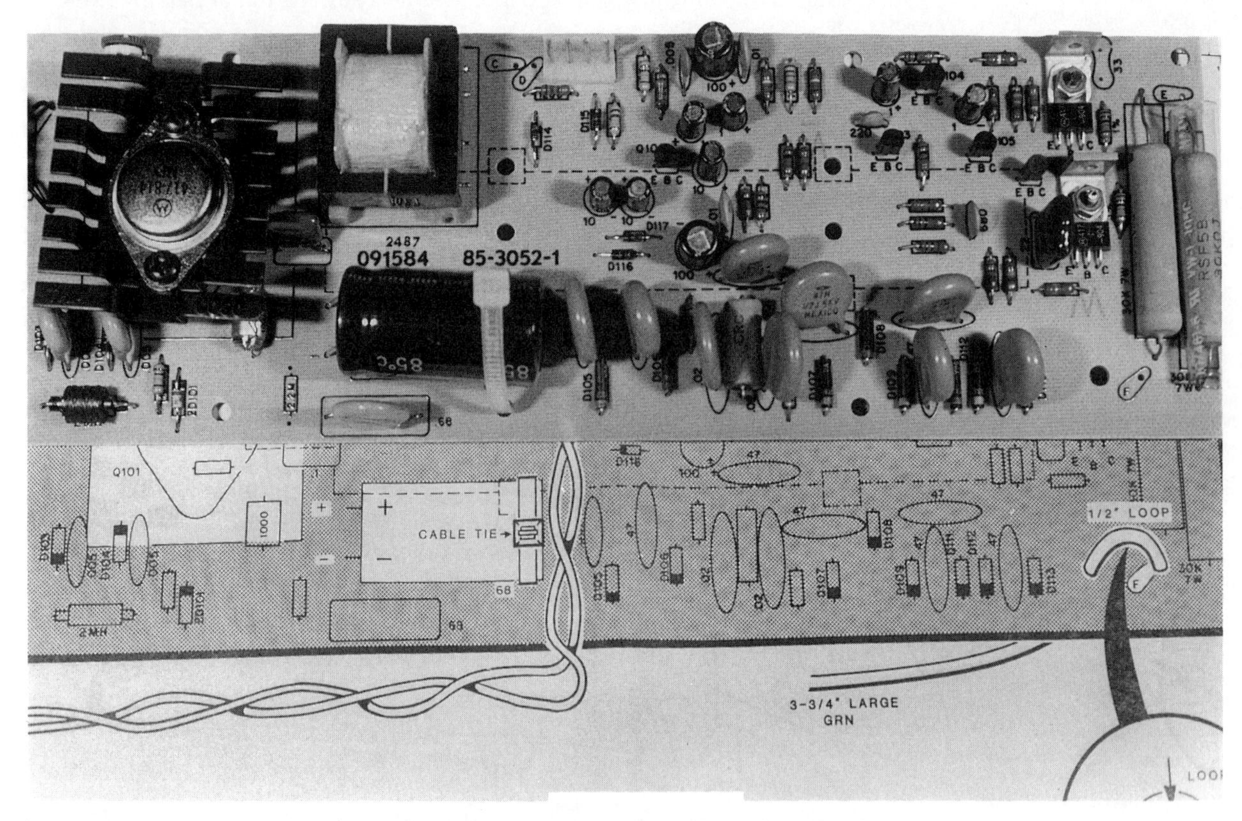

Fig. 8-6. All of the major components have been assembled onto the ET-4200 PCB's top side. Now the board must be flipped over for installing the helium-neon laser tube.

8-8). In this configuration, the laser tube is exposed and extreme caution must be exercised in avoiding the direct exposure of the intense laser light to the human eye.

A small potentiometer (R110) is turned with a provided tool for adjusting the beam's output. The goal during this adjustment phase is to obtain a steady, glowing, evenly illuminated beam of light. Additional fine tuning of this control is possible through using the ETA-4200 Laser Receiver as a supplemental testing device. In this case, a controllable audio signal is sent through the external input jack of the Laser Trainer. This signal is then received over the laser beam by the ETA-4200. Minor adjustments of R110 are then used for obtaining the highest possible audio output from the ETA-4200.

The second adjustment that must be made to the ET-4200 prior to its use is the alignment of the laser beam and the output shutter. The output shutter is used as a safety "trigger" for controlling the output of the laser beam. In order to be effective, this shutter must be adjusted for the maximum output

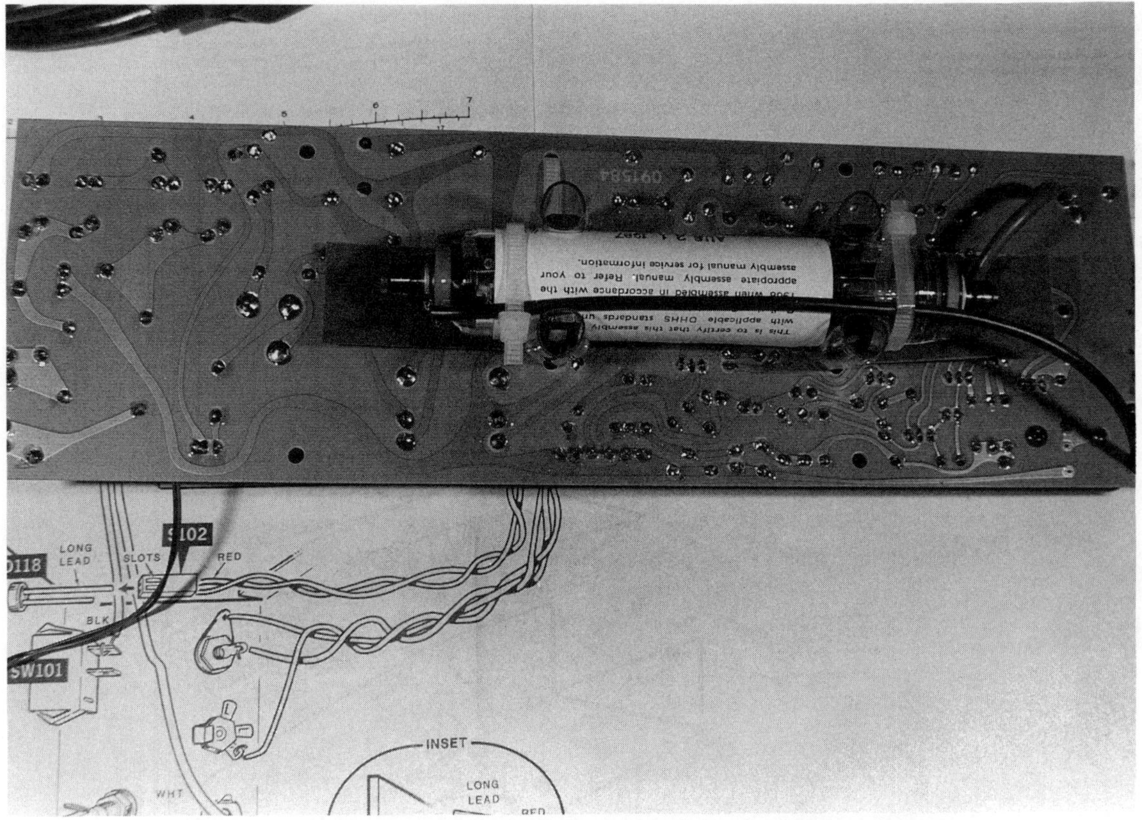

Fig. 8-7. Heavy-duty insulated clips hold the laser tube to the bottom of the Laser Trainer's PCB. Power connections to the tube are made with component-based wires which are fed through holes in the board.

of light through the ET-4200's output aperature. Once again, visual sighting is used during this adjustment. Therefore, extreme care must be exercised in avoiding exposure of the human eye directly to the laser light.

If both of these adjustments have been made properly, the ET-4200 chassis top and bottom can be fixed together and the Laser Trainer safely operated. The average completion time for the assembly of this kit is 2½ hours.

HEATH ETA-4200 LASER RECEIVER KIT

Supporting the ET-4200 Laser Trainer is the Heath ETA-4200 Laser Receiver. This small, battery-powered receiver is specifically designed to receive and decode the modulated beam generated by the ET-4200. A photovoltaic cell acts as the light-sensitive element in the ETA-4200's beam receiving circuit.

Based on the reception of modulated coherent light on this photo cell, a proportional output current is piped through a current-to-voltage converter.

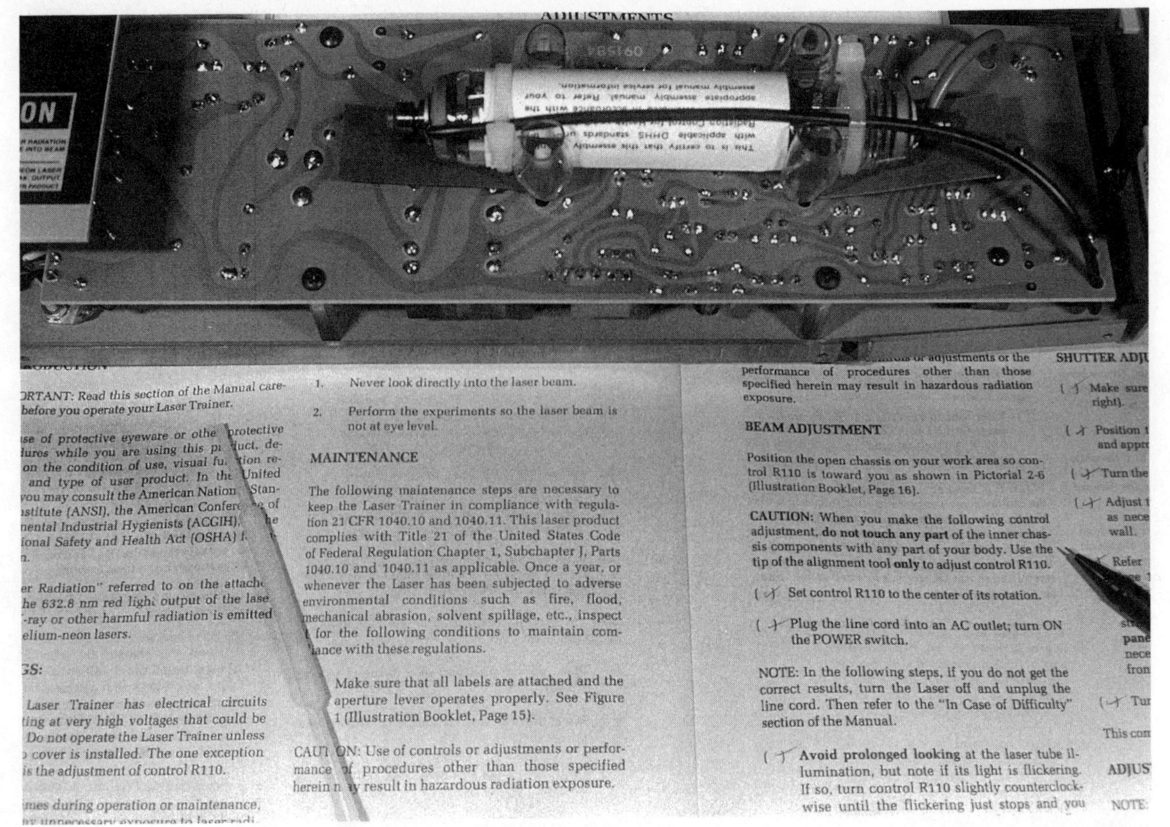

Fig. 8-8. Initial testing of the completed Laser Trainer is performed with the protective top cover removed.

Two limiting potentiometers reduce the effect of ambient light falling on this photo sensor, thereby reducing the ETA-4200's sensitivity to ambient light.

There are two means for interpreting the voltage signal generated by the photo sensor. The first is a signal strength meter. A voltage amplifier drives this meter with the amp's gain controlled through a sensitivity potentiometer. Similarly, an audio amplifier directs the sensor's signal to an internal speaker that provides an adequate interpretation of the received beam modulation.

Like the previously described Laser Trainer, the Laser Receiver kit is supported by excellent documentation (see Fig. 8-9). Guiding the assembly of this kit is an excellent 25-page assembly manual along with a 4-page Illustration Booklet. Beginning with the simple PCB assembly, a total of 19 components (resistors, capacitors, potentiometers, jumpers, and ICs) are soldered into place. Contrary to the Laser Trainer kit, none of these components are supplied on a taped strip. Excellent illustrations and part's descriptions, along with a low part's count, eliminate the need for such a convenience, however.

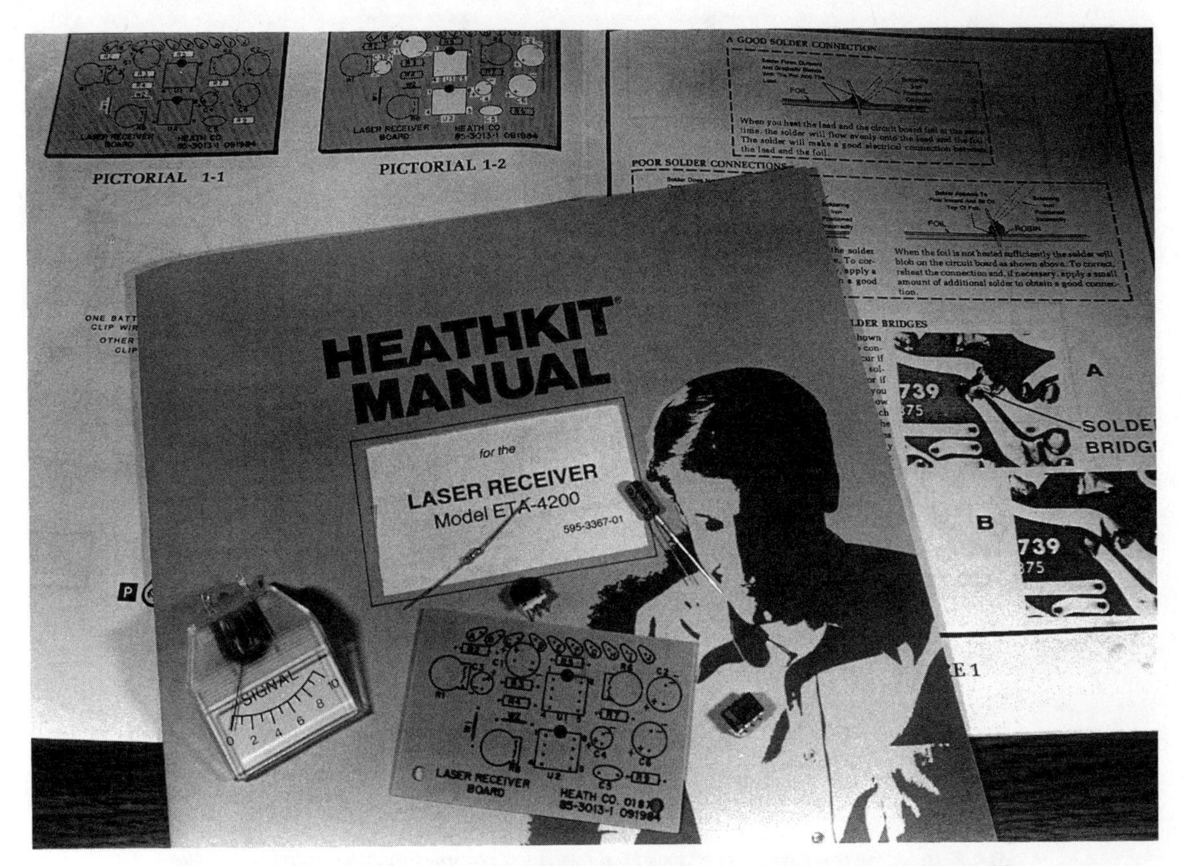

Fig. 8-9. Only a modest number of parts are needed for building Heath's ETA-4200 Laser Receiver.

A total of 10 wires are prepared and soldered into place on the Laser Receiver PCB (see Fig. 8-10). The other ends of these wires all have a final resting spot on the chassis-mounted power switch, meter, sensor, and speaker. In preparation of these final wiring connections, all of the required hardware is mounted to the chassis. Only the assembly of the delicate photo sensor could provide a reasonable pause in the otherwise quick-and-simple assembly of the ETA-4200. Remember, haste makes waste, and failing to observe the sensor's special handling requirements at this point could seriously degrade the overall effectiveness of the final Laser Receiver.

With all of the parts and wires properly soldered into place, the Laser Receiver is now ready for its initial tests. Two 9-volt (1604A) batteries are required for the operation of the ETA-4200. Three controls must be adjusted prior to using the Laser Receiver: null, sensitivity, and volume. During this testing phase the cabinet top is removed. Based on the light produced from a flickering light source (a TV set works great), the wiring of the Laser Receiv-

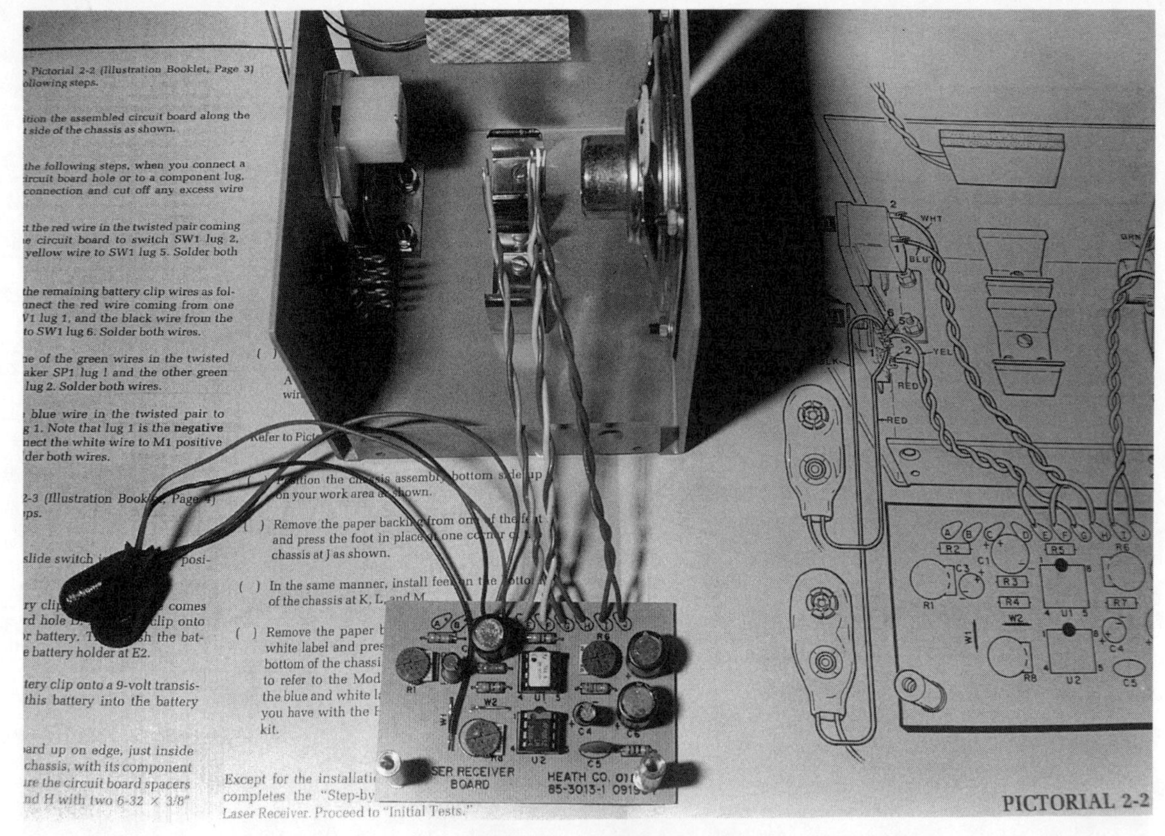

Fig. 8-10. After the Laser Receiver's PCB has been completely assembled, all of the wiring connections are then made between the board and the chassis.

er is rudimentarily examined through meter deflections and speaker output. If these tests are successfully passed, the top is secured into place and the Laser Receiver is ready to reproduce laser beam modulations. The average completion time for the assembly of this kit is 1½ hours.

THE FINAL JUDGMENT

Both the Laser Trainer and the Laser Receiver functioned properly upon their initial testing. During the assembly of each kit, however, several minor errors were noted.

ET-4200:

- Resistor R116 has a value of 1K ohms, but is incorrectly listed as 390 ohms on the PCB silk screen.
- Resistor R135 has a value of 4.7K ohms, but is incorrectly listed as 3.9K ohms on the PCB silk screen and as 6.8K ohms on the schematic diagram.

- The laser tube was missing both of its rubber end caps. This absence increases the amount of laser light inside the Laser Trainer cabinet housing.
- The two ¼-inch Phillips head screws are not pictured in Pictorial 2-5 of the booklet. These screws should be placed in the holes near the rear panel of the Laser Trainer.

ETA-4200:

- Pictorial 2-1 is located on page 3 of the Illustration Booklet, but is incorrectly referenced on page 2.

During operational experiments with the Laser Trainer/Laser Receiver combination, a modulated audio signal was transferred with the laser beam from the ET-4200 to the ETA-4200 (see Fig. 8-11). Although the audio output was not of a high-fidelity quality, the generated sound was recognizable. To ensure success in these modulation experiments, the input signal must be

Fig. 8-11. Introductory laser-based signal modulation studies can be easily performed with the Laser Trainer/Laser Receiver combination.

controllable. Otherwise, the signal can "overdrive" the laser beam, resulting in a distorted audio output.

In support of the experimenter using these two products, Heath markets an educational course dealing with understanding and applying laser technology (Model EE-110). Supplementing the course's 343 page text is a set of 15 different laser-oriented experiments. Each experiment comes complete with all of the parts that are required for building such projects as a laser interferometer for examining the critical angle of a laser beam. Together, this course along with the ET-4200 and ETA-4200 form a valuable, low-cost introduction into laser technology and laser applications.

9

IR Remote Control Systems

An IR remote control system consists of three major components: an emitter, a photodetector, and a communications link. Depending on the desired control application, many different forms of emitters, detectors, and links can be found in an IR remote control system. Generally speaking, however, the two most common control applications for these systems are found in fiber-optic systems and command-oriented, remote-control, function-selection systems. Previous chapters, most notably Chapter 3 and Chapter 5, covered the general theory and operation of a command-oriented system. Therefore, the remainder of this chapter examines in detail IR fiber-optic remote-control systems.

The first element in a fiber-optic system is the emitter. There are two common types of IRED surface emitters: *homojunction* and *heterojunction* LEDs. Homojunction LEDs are the simplest of these fiber-optic emitters. Consisting of a single junction (hence, *homo*junction), these diodes radiate low power photons in a scattering pattern with very little opportunity for control.

Heterojunction emitters, on the other hand, radiate a more controlled beam of light with all photon emissions confined to a reduced area. This beam "focusing" is accomplished through an etched hole placed in the surface of the GaAs substrate. By restricting the current flow to an area in the vicinity of this hole, all of the generated photons will radiate from this controlled opening. Furthermore, a lens is usually fixed over this hole for collimating the photons into a more concentrated beam.

Yet another type of heterojunction LED that makes an excellent single-mode fiber-optic emitter is the injection GaAs laser diode (see Chapter 8). In this diode, the photon emission is generated from the edge of the laser's junction. Although the output power from this emitter is superior to the previously mentioned emitters, a low degree of collimation reduces the performance of the laser diode in multimode fiber-optic systems.

In a fiber-optic system, a photodetector is used for receiving the signal generated by the emitter. Photodiodes (PN and PIN) and phototransistors are the most frequently encountered fiber-optic detectors. Both of these silicon-based detectors have specific strengths and weaknesses when dealing with emitter signals. Generally speaking, photodiodes are used in weak-signal circuits, while the higher noise phototransistor is an ideal detector in microcomputer-based TTL interfaces.

There are five definitions that are used for determining the suitability of a specific detector in a fiber-optic system:

- **Response Time**—The time required by the detector to alter its voltage based on a change in light intensity.
- **Responsivity**—The detector's current production based on the emitter's radiance.
- **Spectral Response**—The responsivity for various wavelengths.
- **Noise**—A signal-degrading factor.
- **Dark Current**—The current output from a non-active detector.

Linking an emitter and a detector together in a fiber-optic system is achieved via a length of light-propagating, optical-glass fiber. This fiber exhibits total internal reflection and continuous refraction over its entire length. Around this glass fiber there is a special material with a lower refractive index known as *cladding*. The purpose behind the lower indexed cladding is that the critical angle for the fiber as a whole is increased. In turn, this heightened critical angle limits the angular distribution of light that can be transmitted through the fiber.

One or more of these cladded glass fibers can be housed within an opaque sheath called a jacket. The jacket is a protective plastic cover which provides a more durable package for manipulating cladded glass fibers.

There are seven factors that govern the application of a fiber-optic system:

- **Numerical Aperture (NA)**—The glass fiber's light-gathering ability. For,

$$\text{Sin } \Theta = \sqrt{n1^2 - n2^2/n0}$$

where,

Sin Θ = maximum angle of acceptance for internal reflection.

n1 = refractive index for the fiber core.
n2 = refractive index for the fiber cladding.
n0 = refractive index for the input medium.

- **Source Area**—The emitter's surface area. This region must be smaller than the fiber's face.
- **Source Radiation Pattern**—The smaller the degree of emitter beam divergence, the better the detector's reception through the fiber cable.
- **NA Coupling Loss**—The difference between the beam divergence of the emitter and the NA of the fiber core.
- **Surface Loss**—The effect of unpolished glass fiber on the emitter source. A major portion of surface loss is due to Fresnel reflection.
- **Attenuation**—The internal loss of energy within the fiber. Absorption by the glass, scattering of photons, and bending all contribute to attenuation.
- **Dispersion**—The internal distortion of the signal within the fiber.

IR EMITTERS AND DETECTORS

Complete IR fiber-optic systems can be built from these IR emitters and detectors. Each individual component is housed within a special case that is compatible with conventional fiber-optic fasteners and connectors. This packaging method aids in minimizing several of the fiber-optic system performance limiting factors.

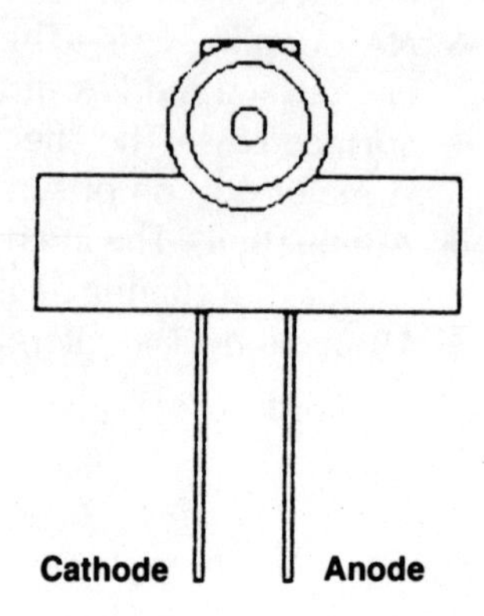

Fig. 9-1. Pin assignments for General Electric GFOE1A1.

Product Example: General Electric GFOE1A1

Package Configuration: AMP Optimate

Reverse Voltage: 6.0 V

Forward Voltage: 1.7 V

Wavelength: 940 nm

Rise Time 0 to 90% of Output: 300 ns

Fall Time 100 to 10% of Output: 200 ns

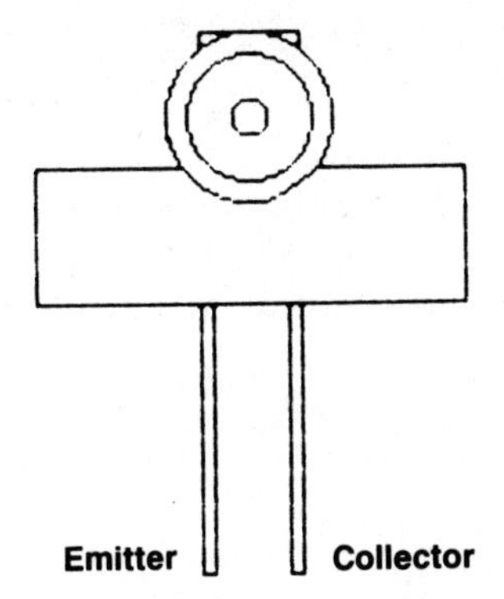

Fig. 9-2. Pin assignments for General Electric GFOD1A1 and GFOD1B1.

Product Example: General Electric GFOD1A1 and GFOD1B1

Package Configuration: AMP Optimate

Collector Dark Current: 100 nA

Responsivity: 70 μA/μW and 1000 μA/μW, respectively

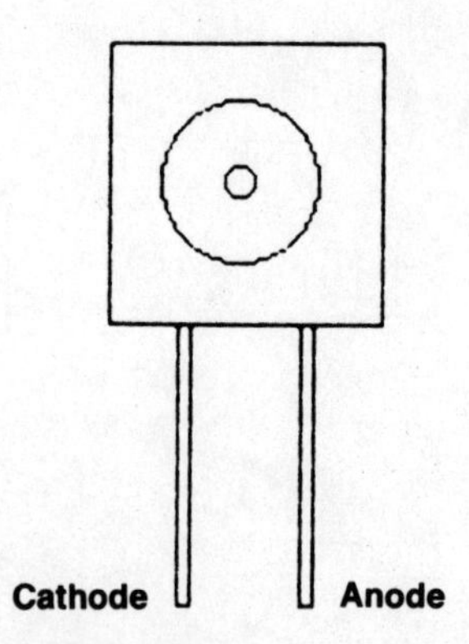

Fig. 9-3. Pin assignments for Motorola MFOE71.

Product Example: Motorola MFOE71

Package Configuration: 363-01 FLCS

Reverse Voltage: 6.0 V

Forward Voltage: 1.5 V

Wavelength: 820 nm

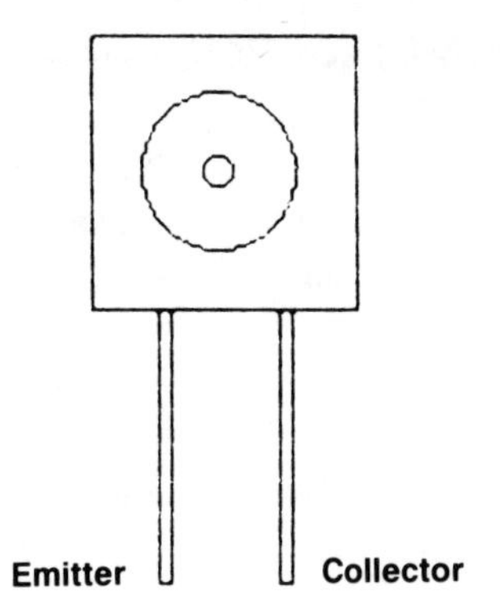

Fig. 9-4. Pin assignments for Motorola MFOD72.

Product Example: Motorola MFOD72

Package Configuration: 363-01 FLCS

Collector Dark Current: 100 nA

Responsivity: 125 μA/μW

FIBER-OPTIC COMMUNICATOR

A fiber-optic communication system offers several advantages over a similar open-air IR counterpart. Chief among these benefits is the reduction in signal attenuation caused by ambient light pollution. Balanced against this contribution, however, is an unfortunate dependence on a fixed length of cable.

Construction Notes

The Fiber-Optic Communicator is a two-part transmitter and receiver circuit (see Fig. 9-5). Each circuit can be constructed on a separate circuit board

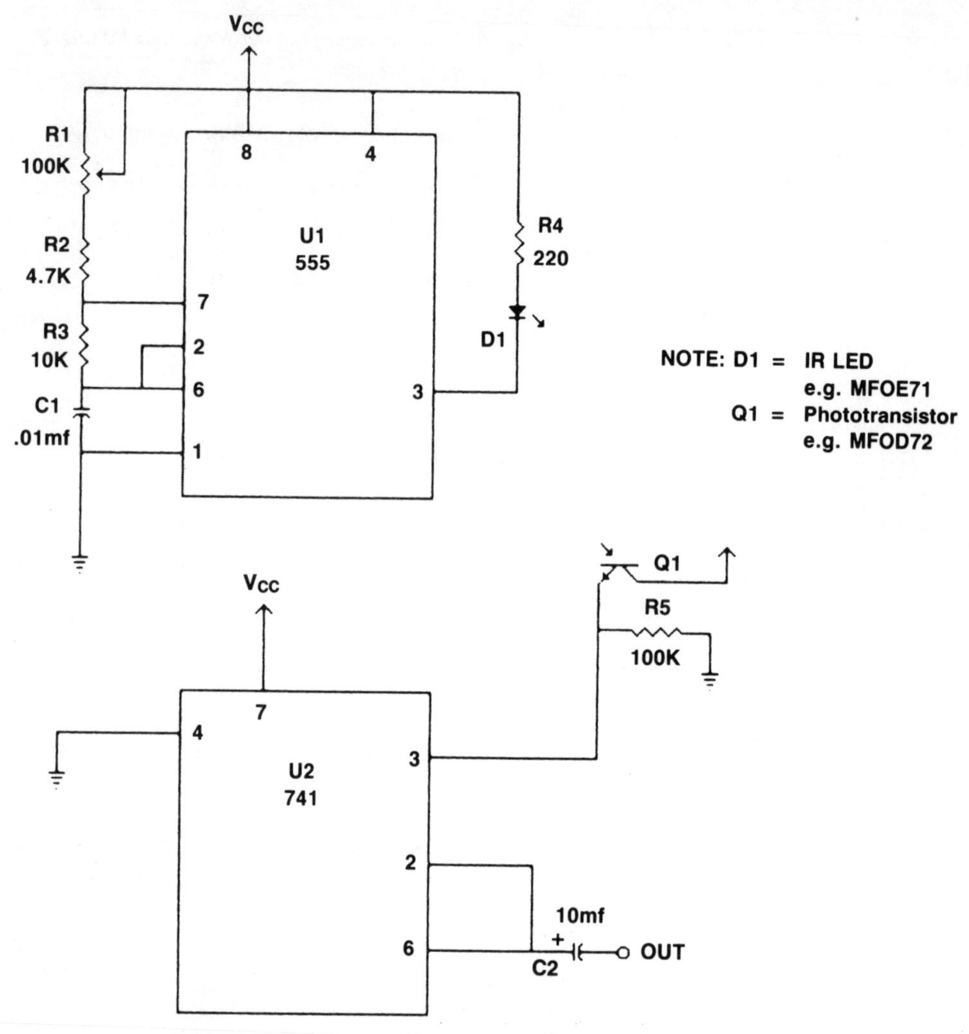

Fig. 9-5. Schematic diagram for Fiber-Optic Communicator.

and linked together via a 5-meter length of fiber-optic cable. Increasing the length of this cable results in signal attenuation and dispersion.

If a suitable amplifier (e.g., based on the LM386 audio amplifier IC) and speaker are connected to the output of the 741 Op Amp (U2), the completed Fiber-Optic Communicator generates a tone during operation. Use potentiometer R1 to adjust the frequency of this tone.

In order to use an optical cable link longer than 5 meters, either an increased power supply or an in-line signal amplifier is necessary. A practical ceiling to increase the current supply to is 100 mA at the collector of the detector. Additionally, the current of the emitter circuit can be increased to a maximum rating of 100 mA. Combining these two current increases raises the practical cable length limit to 15 meters. The better method for increasing the distance between the emitter and detector is to run the fiber link through one or more in-line signal amplifiers. These in-line amplifiers are also known as fiber-optic transceivers.

FIBER-OPTIC TRANSCEIVER

An in-line amplifier or transceiver is an ideal data link between two distant fiber-optic stations. By inserting the transceiver at a halfway point along the data path, the distance between the emitter and the detector can be increased without any appreciable signal degradation. Furthermore, there is no need for juggling the source current to either the emitter or the detector in an attempt to stretch a fiber link. The transceiver is a complete, self-contained signal amplifier that effectively doubles the distance between the source and destination.

Construction Notes

In a digital system, very few components are necessary for designing a transceiver (see Figs. 9-6, 9-7, and 9-8). The Fiber-Optic Transceiver is a low-resolution example of a remote data link that can be applied directly between the emitter and detector pair from the Fiber-Optic Communicator. Based on a CMOS dual D flip-flop (4013; U1), a receiving phototransistor (Q1) is stimulated and subsequently initiates a corresponding pulse on the emitting IRED (D1). The speed of the CMOS logic IC minimizes any noticeable time differential between the reception and transmission of the signal.

An excellent test for the completed Fiber-Optic Transceiver consists of inserting this data link between the emitter and detector from the Fiber-Optic Communicator. During this test, use two separate 5-meter fiber cable lengths with one running from the emitter to the Fiber-Optic Transceiver's phototransistor and the other attached between the transceiver's IRED and the detector. Actually, the result from this elaborate connection scheme differs very little from the direct 5-meter connection between the emitter and the detector—a tone can be heard through the detector's amplifier/speaker

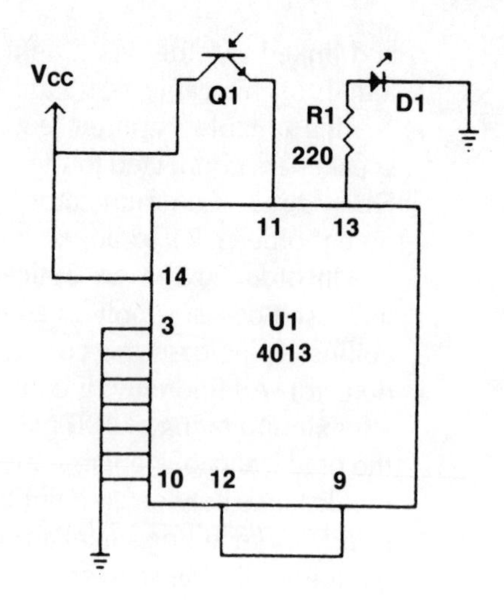

NOTE: D1 = LED
e.g. MFOE71
Q1 = Phototransistor
e.g. MFOD72

Fig. 9-6. Schematic diagram for Fiber-Optic Transceiver.

combination. One major difference between the two connection methods, however, surfaces in the distance achieved through the transceiver connection. Incorporating a transceiver into a fiber-optic system results in a greater distance realization without causing a dangerously high increase in system current.

MOBILE ROOM SCANNER

Coupling the output from an IRED emitter/detector pair to an analog device must be performed via a series of gates, logic converters, pulse extenders, and drive circuits. This complex conversion process is necessitated because

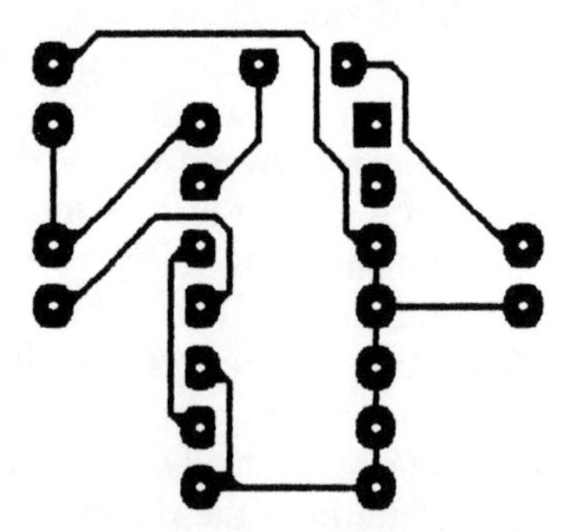

Fig. 9-7. Solder side for the Fiber-Optic Transceiver template, shown at 2X size.

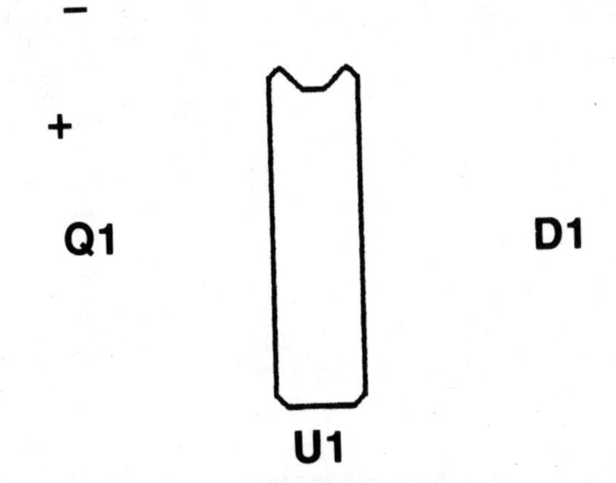

Fig. 9-8. Parts layout for the Fiber-Optic Transceiver PCB.

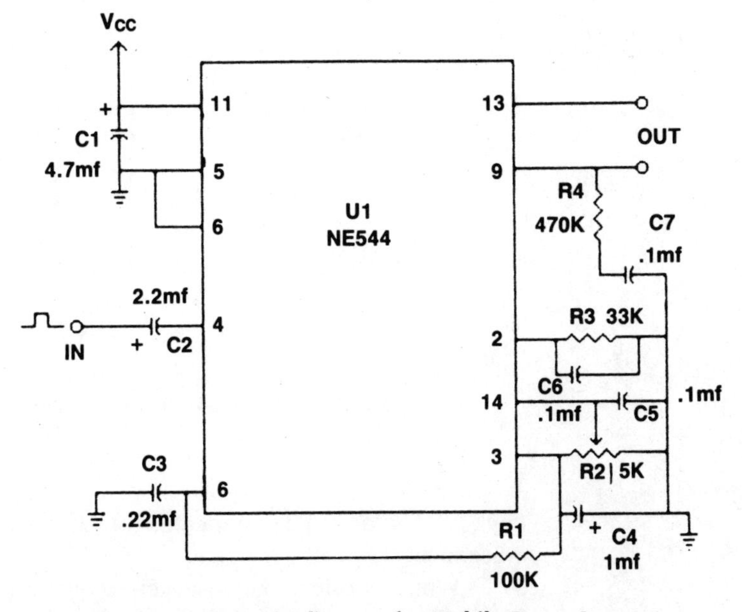

Fig. 9-9. Schematic diagram for Mobile Room Scanner.

the digital output from the IR remote control system is inadequate for directly controlling a high load. Various ICs are available for handling this entire conversion process on a single chip. The Mobile Room Scanner is an example of applying an IR emitter/detector pair to a small dc motor.

Construction Notes

A simple IR emitter/detector system can be directly interfaced to the input of the servo amp IC (see Figs. 9-9, 9-10, and 9-11). Likewise, a small 3.0 Vdc

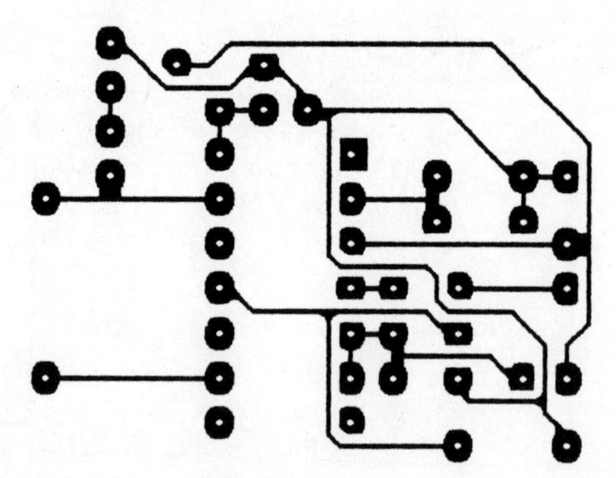

Fig. 9-10. Solder side for the Mobile Room Scanner template, shown at 2X size.

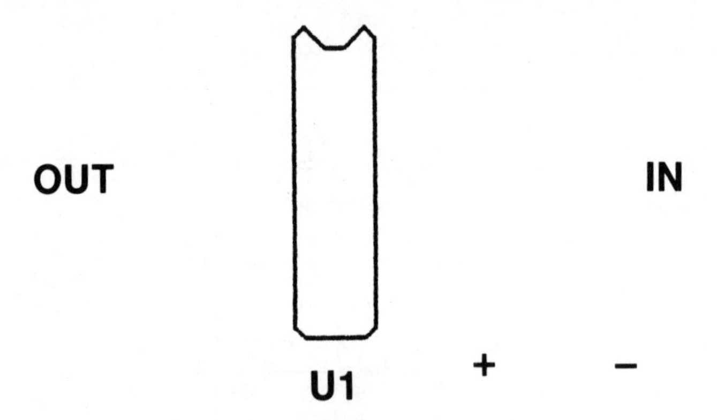

Fig. 9-11. Parts layout for the Mobile Room Scanner PCB.

electric motor can be connected directly to the outputs of the Mobile Room Scanner.

When a positive pulse is received on the input of the NE544, the direction of drive on a connected dc motor is reversed (pins 7 and 11). This action is the prime design feature in the operation of the Mobile Room Scanner. By combining the IR emitter/detector pair, servo amp circuit board, and dc motor together onto a singular chassis, the Mobile Room Scanner can be made to continually survey a room for objects.

One application for this chassis combination is in replacing the radio control circuitry from a commercial toy with the Mobile Room Scanner circuitry. This conversion eliminates the need for activating a control box for controlling the direction of travel for the toy. In other words, a small electric car will be able to drive throughout a room and avoid hitting any obstacle in its path.

REMOTE CONTROL

Another application for the IR emitter/detector pair is in IR remote control systems. A popular example of this application is in the remote channel selectors used in television sets. These IR remote control systems rely on a controlled pulse from the emitter that is sensed and decoded by the detector. Once again, special ICs are available for decoding the detector received pulses.

Construction Notes

A different means of detection must be used in television IR remote control systems (see Fig. 9-12). In this case, an IR PIN photodiode (D1) is used for receiving the emitter's pulse. Decoding this reception is left to a dedicated remote control IC.

The Texas Instruments SN76832AN (U1) is an example of a common IR remote control system decoder. A simple 40 kHz pulse transmitter based on

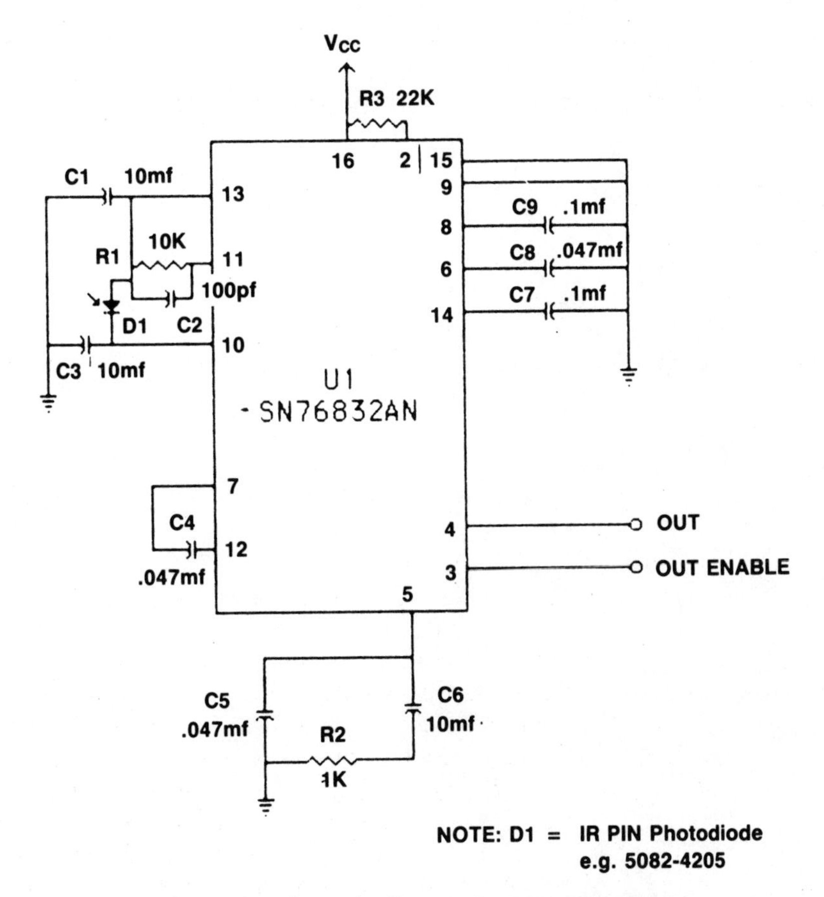

Fig. 9-12. Schematic diagram for Remote Control.

an IR emitter can be used for sending a one-channel signal to this receiver/decoder. More complex, multiple-channel signals, however, must be transmitted via a special transmitter IC. The Motorola IR transmitter/receiver IC pair XC14457 and XC14458, respectively, can control up to 16 channels through single-entry commands. Additionally, up to 256 channels can be addressed with double-entry commands. Combining these sophisticated IR transmission/reception encoders/decoders together with IR emitter/detector pairs produces a signal density that is capable of performing numerous complex household tasks—all at the touch of a button.

10

Digital GaAs ICs

Based on the relatively "light" weight of gallium arsenide electrons, extremely "fast" electronic circuits can be designed that exhibit a six-fold speed increase over comparable silicon packages. Derived from high-quality precipitations in Liquid Encapsulated Czochralsky (LEC) crystal growing techiques, GaAs FETs exhibit electron velocities in the neighborhood of 5×10^7 meters per second. By further combining the insulative features of the GaAs substrate, performance-destroying parasitic capacitances can be similarly minimized. The end result is a high-speed, high-quality, radiation-hardened, temperature-stable circuit that rivals the performance specifications that can be obtained with tempermental cryogenic Josephson-Junction devices (also known as *superconductors*).

Gallium arsenide ICs can be formed from a number of different logic families that are unique to high-speed devices: D-MESFET, E-MESFET, and HEMT. The most predominant of these various GaAs logic families is the D-MESFET. The construction of the D-MESFET serves as an excellent illustration of the performance benefits achieved through GaAs fabrication techniques.

The D-MESFET consists of a 100 to 200 nm deep n-type implantation with a 3 to 4 micron separation between the source and the drain. In this configuration, the D-MESFET is similar to a conventional silicon JFET. A significant difference between these two technologies, however, is the high transconductance and low input capacitance of the GaAs D-MESFET. The combination of these two features lends the GaAs FET a bandwidth of 15 to 20 GHz with switching speeds of 50 ps (picoseconds).

Various logic circuits can be designed from the D-MESFET beginning. RAM, BFL (*b*uffered *F*ET *l*ogic), CCFL (capacitor-coupled *F*ET *l*ogic), SCL (*s*ource-coupled *l*ogic), and SDFL (Schottky Diode FET Logic) have all been built from D-MESFETs. Each of these individual circuits exhibit the same performance benefits that are found in the individual D-MESFET. For example, a 256 × 4-bit GaAs RAM possesses an access time of 1 ns with a 2.5 ns cycle time. Unfortunately, all is not bliss in this processing-speed paradise. Operating at this blazing speed does present several design problems concerning the physical layout of the RAM chip carrier, namely lead length, noise, and crosstalk. In an effort to eliminate these three performance-robbing problems, a lead-less chip carrier answers all three of these issues. As described, this GaAs static RAM IC (12G014) is manufactured by GigaBit Logic (a competitor to the GigaBit Logic RAM is the Fujitsu 256 × 4-bit RAM based on D-MESFET and E-MESFET technologies with an access time of 4 ns).

GigaBit Logic's 12G014 256 × 4-Bit Registered, Self-Timed Static RAM (see Fig. 10-1) is used primarily as a high-speed memory cache and memory buffer in memory-intensive computer systems. With equal read-and-write

Photograph courtesy of GigaBit Logic, Inc.

Fig. 10-1. GigaBit Logic's 12G014 256 × 4-Bit Registered, Self-Timed Static RAM. This GaAs digital memory IC exhibits a 2.5 nS read and write cycle time.

cycle times (2.5 nS or 3.5 nS), this NanoRam family member couples its RAM capabilities with latched inputs and outputs, an internally generated write pulse, and either a single-ended or differential clock input. Furthermore, all input and output levels are both ECL and PicoLogic family (another line of popular GigaBit Logic GaAs digital ICs) compatible. This results in a flexible, fully registered, clocked, static RAM IC.

In spite of these performance advantages, GaAs D-MESFETs do suffer from two limitations. First, they require two power supplies for successful operation. Second, a voltage-level-shifting function must be designed into the logic gates for supplying the positive drain voltage's negative switching voltage. Both of these minor limitations have been eliminated by E-MESFET technology.

By using a positive threshold voltage, E-MESFETs are able to avoid the dependence on the dual power supply and voltage-level-shifting function. Unfortunately, along with these benefits there is a new group of performance barriers. Basically, E-MESFETs require an extremely narrow voltage logic deflection (0.5 V) that limits their use in logic circuit construction.

Constructed from molecular-beam epitaxy, the HEMT provides better performance than the E-MESFET while minimizing many of its design

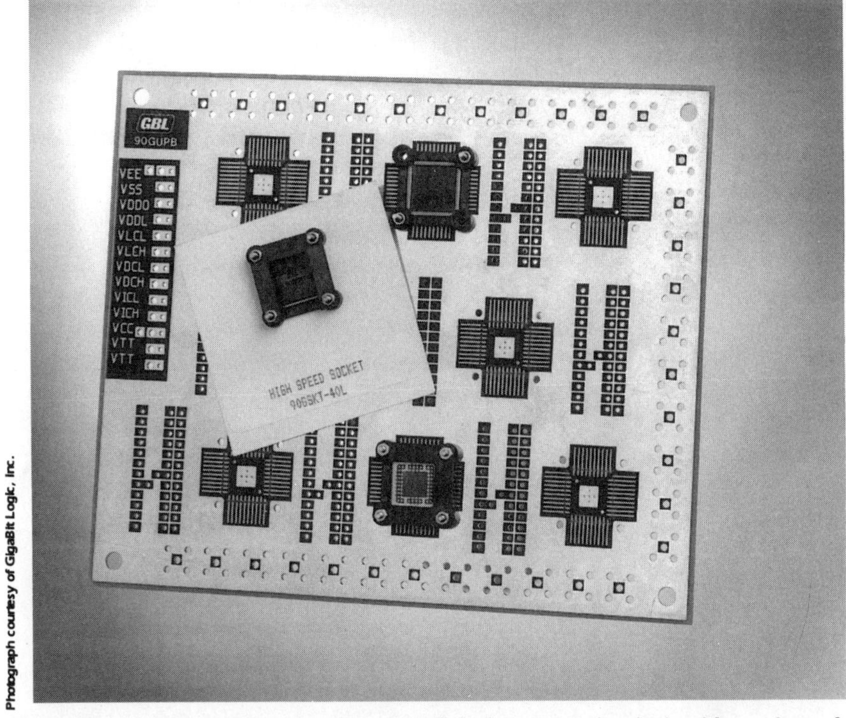

Photograph courtesy of GigaBit Logic, Inc.

Fig. 10-2. An efficient means for designing digital GaAs IC circuits is with a universal prototyping board like this GigaBit Logic example. Up to eight GigaBit Logic ICs with 36 or 40 I/O parts can be interfaced on this board.

restrictions. Although derived from the E-MESFET fabrication process, the HEMT's gate begins to conduct at 1.2 V. This increased voltage threshold is made possible by a heightened electron mobility in the undoped HEMT GaAs channel. Therefore, complete transconductance is possible with voltages near the threshold voltage. The result is a suitable GaAs technology for fabricating complex logic circuits that demonstrate high clock speeds at low temperatures.

Designing with GaAs digital ICs closely follows the same precautions and restrictions that are found in ECL designs.

- Observe ECL and CMOS electrostatic handling precautions with GaAs digital ICs.
- Decouple all power supply pins with capacitors.
- Only glass epoxy boards should be used as a circuit substrate.
- All unused GaAs digital IC inputs must be tied to the power supply. (This is in contrast to ECL designs.)
- All unused GaAs digital IC outputs should be left open.

In addition to being a leading supplier of GaAs digital ICs, GigaBit Logic also manufactures a Universal Prototyping Kit (90GUPK; see Fig. 10-2) that can interface up to eight GaAs ICs into a test circuit. The major component of this kit is the four-layer 90GUPB PCB. The 90GUPB consists of eight gold-plated IC solder sites along with a dedicated dc voltage bus for each layer. One important design consideration of the 90GUPB is that each of these IC solder sites is capable of holding any GigaBit Logic 36- or 40-pin I/O package.

DIGITAL ICs

More manufacturers are quickly entering the race to fabricate digital GaAs ICs. Two manufacturers who are representative of this group are GigaBit Logic and Harris Semiconductor. A complete line of logic and memory ICs are available from these manufacturers featuring fast clock speeds, rapid output transition times, enhanced radiation hardening, extended operating temperature range, and moderate power requirements.

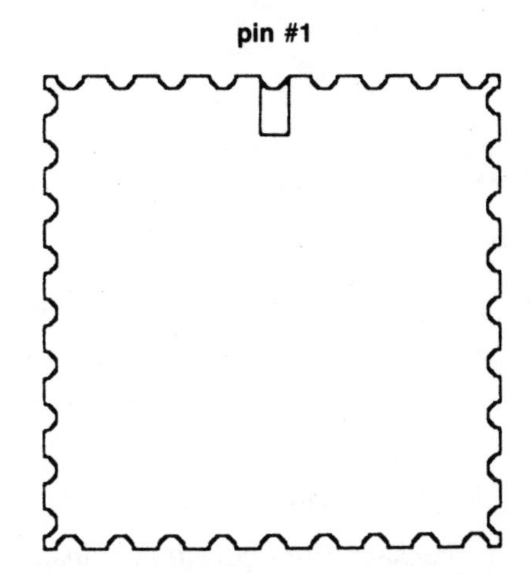

Fig. 10-3. Pin configuration for GigaBit Logic 10G000A and 10G021A.

Product Example: GigaBit Logic 10G000A and 10G021A

Package Configuration: 40 I/O C-leadless CC

Internal Logic Arrangement: Quad 3-input NOR gate and dual-precision D flip-flop respectively

Operating Speed: 290 ps gate delay and 2.7 GHz clock rate, respectively

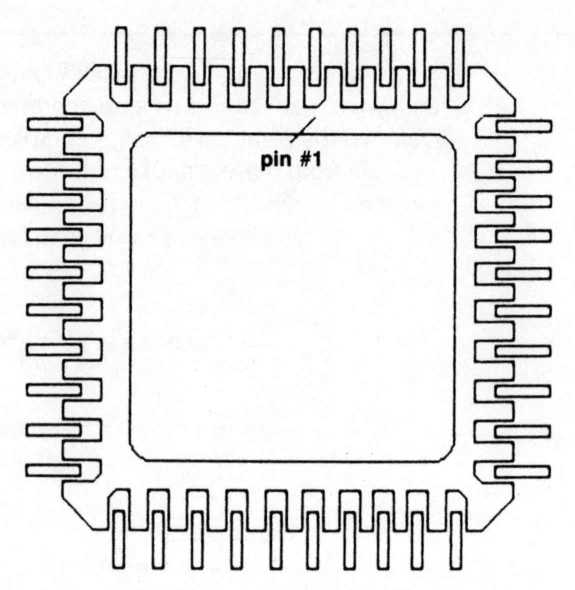

Fig. 10-4. Pin configuration for GigaBit Logic 10G004 and 12G014.

Product Example: GigaBit Logic 10G004 and 12G014

Package Configuration: C-leadless CC and leadless CC, respectively

Internal Logic Arrangement: Quad 2:1 multiplexer and 256 × 4-bit static RAM

Operating Speed: 3.6 Gbit/sec data rate and 2.5 ns cycle time

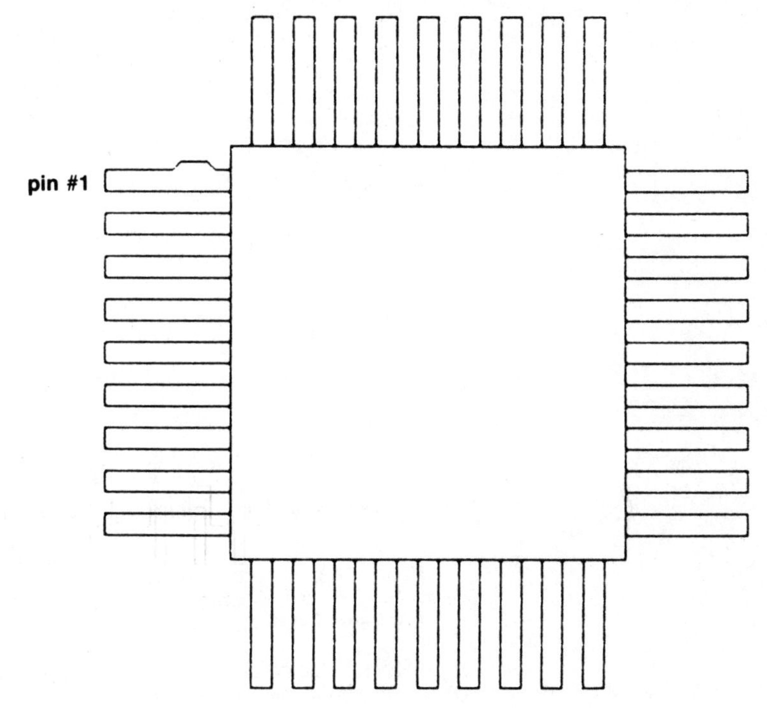

Fig. 10-5. Pin configuration for GigaBit Logic 10G060.

Product Example: GigaBit Logic 10G060

Package Configuration: 36 I/O flatpack

Internal Logic Arrangement: 2-Stage ripple counter/divider

Operating Speed: 3.0 GHz clock rate

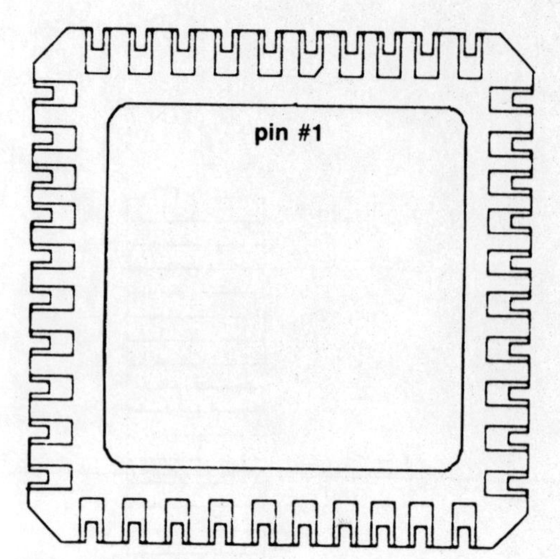

Fig. 10-6. Pin configuration for GigaBit Logic 16G044.

Product Example: GigaBit Logic 16G044

Package Configuration: Dice

Internal Logic Arrangement: Phase/frequency comparator

Operating Speed: 1 GHz input frequency

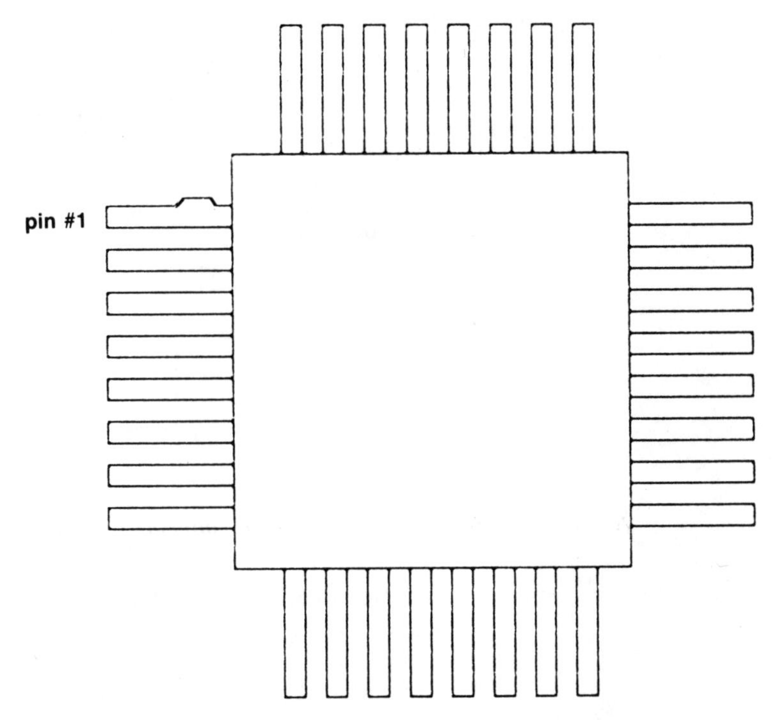

Fig. 10-7. Pin configuration for Harris Semiconductor HMD-12141-1 and HMD-11016-1.

Product Example: Harris Semiconductor HMD-12141-1 and HMD-11016-1

Package Configuration: MSI

Internal Logic Arrangement: 4-Bit universal shift register and divide by 2/4/8, high-speed synchronous counter, respectively

Operating Speed: 1.3 GHz clock speed and 2.0 GHz data input rate, respectively

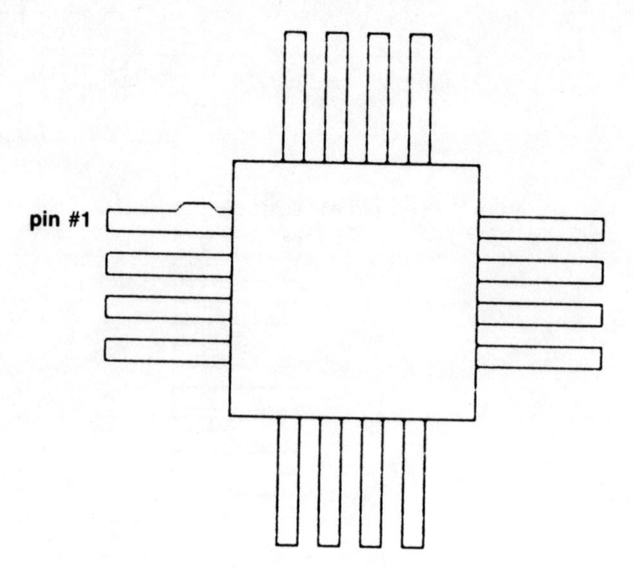

Fig. 10-8. Pin configuration for Harris Semiconductor HMD-11104-2.

Product Example: Harris Semiconductor HMD-11104-2

Package Configuration: SSI

Internal Logic Arrangement: 5-input NAND/AND gate

Operating Speed: 2.5 GHz data input rate

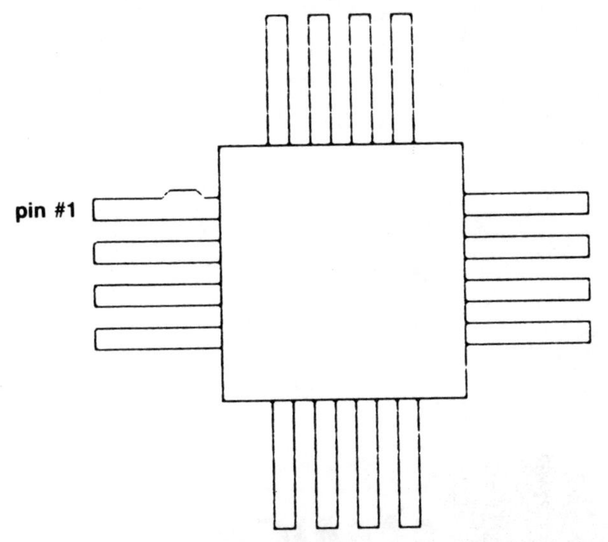

Fig. 10-9. Pin configuration for Harris Semiconductor HMD-11113-2.

Product Example: Harris Semiconductor HMD-11113-2

Package Configuration: SSI

Internal Logic Arrangement: Dual 2-Input exclusive OR gate

Operating Speed: 2.5 GHz data input rate

11

GaAs MMICs

Gallium arsenide *m*onolithic *m*icrowave *i*ntegrated *c*ircuits (MMICs) are the technological flip side to GaAs digital ICs. Possessing attributes similar to analog feedback and amplification ICs, the GaAs MMIC is used extensively in broadband amplifiers for rf and microwave communications. The construction of the GaAs MMIC helps in labeling the differences between these two gallium arsenide technologies.

GaAs MMICs are fabricated with 0.5 micron gates. Conversely, GaAs digital ICs have 1 micron gates. Additionally, GaAs MMICs are designed from FETs or Schottky diodes built from GaAs epitaxial layers. GaAs MMICs also contain thin-film resistors and capacitors on their substrate wafer. It is the presence of these formerly discrete support components within the MMIC that causes the majority of dichotomous identification, as well as design problems. A strict adherence to monitoring epitaxial parameters, FET (or Schottky diode) parameters, and the passive support component parameters during the fabrication process ensures the quality of the final MMIC. Therefore, the key design element separating GaAs digital ICs from GaAs MMICs is the restrictive gate widths and the need for embedded passive elements.

Due to this complexity in design, GaAs MMICs must undergo rigorous testing during virtually every phase of fabrication. Two of the more common tests include an rf test and a dc test. Each of these tests is conducted at both the wafer- and the package-stage of fabrication.

Rf Tests. A typical rf test and its associated parameters would be:

- Testing frequency range 3 to 26.5 GHz.
- A wafer-stage gain of 2 dB.
- A wafer-stage phase difference of 3 degrees.

Dc Tests. A typical dc test and its associated parameters consists of:

- Determine broken gates.
- Locate weak short circuits.
- Testing frequency range 3 to 6 GHz.

The conclusion of these tests yields a MMIC that has been summarily dc-screened and limitedly rf-screened. Of these two tests, the dc tests are more critical. A reliable dc test will determine the threshold voltage for the MMIC's individual gates, along with the gate-to-source capacitance. Therefore, the minimal rf testing is considered insignificant in the fabrication of high-performance MMICs.

One contributing factor to the reduced employment of adequate rf testing is the manufacturer's inability to execute the test during the wafer-stage of fabrication. Ideally, locating a "bad" MMIC at this stage would reduce manufacturing costs and ensure a quality product. Recent developments in test probes by Cascade Microtech, Inc. have produced a test station for automatically evaluating the performance of a MMIC while still encapsulated in a wafer form. The Model 42 Microwave R&D Probe Station tests MMICs with frequencies up to 26.5 GHz directly in the wafer without a supplementary hardware carrier. Discovering faulty MMICs at this stage prevents the wasteful placement of the MMIC in a carrier with the savings from this prevention being roughly equal to the production costs for the MMIC itself.

SIGNIFICANT GaAs MMIC PRODUCTS

Company: General Electric is designing MMICs for military applications.

Operating Frequency: 500 MHz to 94 GHz

Principal Technology: HEMT

Applications:
- Space-based communications
- SDI (Strategic Defense Initiative) Target Acquisition and Control
- Circuitry

Company: Honeywell is designing MMICs for space research.

Operating Frequency: 4 GHz

Principal Technology: Unbiased FET

Applications:
- Space-based communications
- NASA communications systems
- Avionics

Company: Rockwell is designing MMICs for USAF research.

Operating Frequency: 20 GHz

Principal Technology: Unspecified

Company: Texas Instruments is designing MMICs for military defense systems.

Operating Frequency: 6 to 20 GHz

Principal Technology: FET

Applications:
- Communications
- Avionics

Company: Westinghouse is designing MMICs for joint USN and USAF research.

Operating Frequency: 8 to 12 GHz

Principal Technology: Selective ion-implantation FET

Applications:
- ECM (Electronic CounterMeasures) warfare
- Communications
- Avionics

Appendices

Appendix A

Building a GaAs Project

Building your own electronic circuit is an exciting project that has been carefully detailed in this book's preceding chapters. By combining a few dollars worth of parts with a circuit's schematic diagram, an intelligent design is born. Unfortunately, several factors are bound to block the successful implementation of this project. There are financial, mental, and physical reasons that might prevent the introduction of PCBs. While I can't satisfy either your financial or physical difficulties, I can try to lessen the severity of your naivete to electronics construction techniques.

Whether you are a seasoned electronic project builder, or a casual computer user, you need to learn a few construction basics. Most of these building techniques center around the most effective means for translating a circuit from paper into a wired, operating unit. Two specific areas in which this construction education is stressed are the manner in which the circuit is built and the enclosure in which it is placed.

Seasoned electronics project builders may be experienced hands at soldering and printed circuit board etching (if you aren't, see later in this appendix), but use of the E-Z Circuit boards is one method of project wiring that is available to all levels of project builders. Circuits are easily transferred from the printed schematic onto an E-Z Circuit board by using easily cut dry-transfer shapes and pad soldering. Circuit builders wishing more stability for their projects can quickly translate the E-Z Circuit designs onto an etched PCB.

Once you have finalized your design, it is time to place your circuit inside a housing. As a rule, a housing is only necessary for stand-alone and parallel-

or serial-port connection computer-based projects. For example, if your final project is for an internal expansion slot of a personal microcomputer, then you will not need to worry about a housing. Conversely, if your circuit uses a parallel port for interfacing with microcomputers, then you need to consider the design of a circuit housing. In some ways, the familiar, boxy, metal or plastic electronics project cabinet has turned into an antiquated relic. For the majority of the constructed projects, however, the purchased storage cabinet is the ideal project housing solution.

IF IT'S E-Z, IT MUST BE EASY

Custom circuit boards used to be the exclusive domain of the acid-etched PCB. Now Bishop Graphics has introduced a revolutionary concept that could become the next dominate circuit board construction technique. E-Z Circuit (that's what Bishop Graphics calls it) takes a standard, pre-drilled universal PC board and lets the builder determine all of the tracing and pad placements. A special adhesive copper tape is the secret to this easy miracle in PCB fabrication.

The E-Z Circuit system is both an extensive set of these special copper tape patterns and several blank universal PC boards. The blank boards serve as the mounting medium for receiving the copper patterns. There are general-purpose blank boards (Bishop Graphics' #EZ7402 and #EZ7475) and Apple II family blank PC boards (Bishop Graphics' #EZ7464). Each of these boards is made from a high-quality glass epoxy and pre-drilled with IC-spaced holes. There is also a blank edge-connector region on each board for attaching one of E-Z Circuit's copper edge connector strips.

Complementing the blank boards is a complete set of copper patterns. Each pattern is supplied with a special adhesive that permits minor repositioning but holds firmly once its position is determined. This adhesive also has heat resistant properties that enable direct soldering contact with the pattern. Only an extensive selection of patterns and sizes would make an E-Z Circuit design worthy of consideration for project construction. Once again, E-Z Circuit satisfies all of these requirements with edge connector, IC package, terminal, test point strip, tracing, donut pad, elbow, TO-5, power and ground strip, and power transistor patterns. Additionally, each of these patterns is available in several sizes, shapes, and diameters.

Only four simple steps are needed in the construction of an E-Z Circuit project. In step one, you determine which pattern you need and prepare it for positioning. Step two is for placing the selected pattern on the blank PC board. This is a simple process that involves the removal of a flexible release layer from the back of the copper-clad pattern. During step three you insert all of the components into their respective holes. Finally, in step four, you solder the component's leads to the copper pattern. This step should be treated just like soldering a conventional copper pad or tracing. If you use a reasonable soldering iron temperature (anywhere from 400 to 600 degrees Fahrenheit) you won't need to worry about destroying the adhesive layer and ruining the copper pattern.

Making a slight digression on the issue of soldering irons, be sure to use an iron with a rating of 15 to 25 watts. Soldering irons that are matched to the demands of working with E-Z Circuit include: ISOTIP 7800, 7700 (the specifications for these two irons border on the stated E-Z Circuit requirements, but they will work) and 7240, UNGAR SYSTEM 9000, and WELLER EC2000.

The result of this four-step process is a completed project in less time than a comparably prepared acid-etched PCB. Incidentally, the cost of preparing one project with E-Z Circuit versus the etched route is far less. Of course, this cost difference is skewed in the other direction when you need to make more than one PCB, because E-Z Circuit is geared for one-shot production and not assembly line production. The bottom line is, before you decide on your circuit board construction technique (etched PCB or E-Z Circuit), read the remainder of this appendix for the latest advances in acid-etched PCB design. Then evaluate your needs and resources and get to work on your first powerful project.

TAMING THE SOLDER RIVER

Before the first bit of solder is liquefied on your PC board, an overall concept must be organized for fitting the completed project into an enclosure. Small enclosures are usually preferred over the larger and bulkier cabinets simply because of their low profile and discreet appearance. Space is limited within the narrow confines of such a sloped enclosure, however.

It's quite easy to envision a two-dimensional circuit schematic diagram and then forget that the finished, hard-wired circuit will actually occupy three dimensions. Capacitors, resistors, and ICs all give a considerable amount of depth to a finished project PC board. Fortunately, the effects of these tall components can be minimized through some clever assembly techniques.

With only a few exceptions, all components must be soldered to a PC board as closely as possible. One exception to this rule is in leaving adequate jumper wire lengths for external enclosure-mounted components, such as switches and speakers.

Both component selection and their mounting methods directly affect a project's PC board depth. For example, the selection of a horizontally oriented, miniature, PC-mountable potentiometer over a standard vertically oriented potentiometer can save up to ½ inch off of a board's final height. Likewise, flat, rectangular metal-film capacitors offer a space savings whenever capacitors of their value are required (usually .01 μF to 1.0 μF).

If a disc or monolithic capacitor is used on a circuit board, the lead can be bent so that the capacitor lies nearly flat against the PC board. The capacitor can be pre-fitted before soldering it into place, and the required bends can be made with a pair of needle-nosed pliers. Be sure that the leads of every component are slipped through the PC board's holes as far as possible before soldering them into place. Excessive leads can be clipped from the back side of the board *after* the solder connection has been made. At this point, a precautionary note pertaining to overly zealous board compacting is necessary. Do not condense a board's components so tightly that undesired leads might touch; a short circuit is the inevitable and unwanted result for this carelessness. Also, some components emit heat during operation. Therefore, component spacing is mandatory for proper ventilation.

One way to minimize PC board component crowding is by using a special mounting technique with resistors and diodes. The common practice for mounting resistors and diodes is to lay them flat against the PC board. This technique is impractical, and occasionally impossible, on the previously described PC boards. A superior technique is to stand the resistor or diode on its end and fold one lead down until it is parallel with the other lead. The component leads can then be placed in virtually adjacent holes.

One final low-profile component that is an absolute necessity on any PC board that uses ICs is the IC socket. An IC socket is soldered to the PC board to hold an IC. Therefore, the IC is free to be inserted into or extracted from the socket at any time. Acting as a safety measure, the IC socket prevents any damage that might be caused to a chip by an excessively hot soldering iron if the IC were soldered directly to the PC board. IC chips are extremely delicate and both heat and static electricity will damage them. After the project board has been completely soldered, the ICs are finally added.

THE FINISHING TOUCH

The most easily acquired enclosure for your finished project is a metal or plastic cabinet that can be purchased from your local electronics store. Radio Shack makes a stylish, wedge-shaped enclosure (Radio Shack #270-282) and a two-tone cabinet (Radio Shack #270-272 and #270-274), all three of which are perfect for holding your project. Another slightly less attractive but still useful project enclosure that is also available from Radio Shack is the Experimenter Box (Radio Shack #270-230 through #270-233 and #270-627). The Experimenter Box comes in five different sizes for holding any size of project.

If none of these enclosures fit your requirements, you can also build your own project cabinet. The best material for building your own enclosure is with one of the numerous, inexpensive types of sheet plastics that are currently available. Materials such as Plexiglas are easy to manipulate with the right tools and adhesives. Plexiglas sheeting can be cut with a hand saw or a power jig saw. Leave the protective paper covering on the Plexiglas while cutting to prevent it from splitting or becoming scratched.

Construct all sides of the enclosure by joining the sides together and laying a thin bead of a liquid adhesive along the joint. A powerful cyanoacrylate adhesive, such as Satellite City's SUPER "T" will bond two pieces of Plexiglas together immediately. Be sure that all panel edges are perfectly aligned before applying the adhesive. One side of the final enclosure design should not be joined with the adhesive. This panel should be held on with screws so that it is easily removed for future access to your circuit.

Hopefully you have finished reading this portion of Appendix A before you have started your project's actual construction. If so, good; you have saved yourself several hours of headaches and quite probably several dollars in wasted expenses. If not, all is not lost. Just review your current construction in light of what you have learned in this appendix and make any needed changes in your construction procedures. At least when you make your next project, you will know all of the secrets to successful circuit sustentation.

PCB DESIGN WITH CAD SOFTWARE

For the most part, if you own either an Apple Macintosh or an IBM PC family microcomputer, then you have the potential for designing your own printed circuit boards. This potential is only realized after the purchase of some specialized design software, however. A rather general term is applied to this type of software—Computer-aided design, or CAD software. CAD is a relatively young field with true dominance already present in the IBM PC arena. This definitive CAD software product is called AutoCAD 2 and it is manufactured by Autodesk, Inc. Don't be fooled by other CAD products that are designed for IBM PCs and their clones and cost less than AutoCAD's roughly $2000 price tag. These cheaper programs are totally inferior to AutoCAD. Without question, the AutoCAD environment is the most powerful CAD software that is currently available for microcomputers and yet it remains flexible to every users' demand.

One fault that is frequently attached to CAD software like AutoCAD is the lack of a dedicated PCB template formulation application. In other words, programs, like AutoCAD, must be customized with user-created symbol libraries before PCB templates can be expertly designed in a minimal amount of time. If you think that there's got to be a better way, then let Wintek Corporation show you the path to this ideal solution. Two professional pieces of CAD software, smARTWORK and HiWIRE, are vertical applications with specific PCB template fabrication virtues. These programs sport an impressive list of PCB template design features:

- a silkscreen layer
- text lettering for every layer
- variable trace width
- user-definable symbol libraries
- automatic solder mask generation
- 2X check prints
- high-quality dot-matrix 2X artwork

- numerous printer/plotter drivers
- rapid printouts
- AutoCAD *.DXF file generation

Making It with smARTWORK

In order to provide a clearer picture of the operation of smARTWORK, Figs. A-1 through A-4 illustrate the major steps involved in the design of a PCB. This representative design is the template for the Digital Timer project from Chapter 1.

The bottom line for these Wintek Corporation PCB CAD products is that they will aid you in the design of custom PCB templates for less than half the cost of AutoCAD and with double the creative power.

Making It with Quik Circuit

Until recently, your only PCB CAD route was with the previously described smARTWORK. The cost in the IBM hardware alone was prohibitive to some designers. Most of this need for costly hardware changed with the

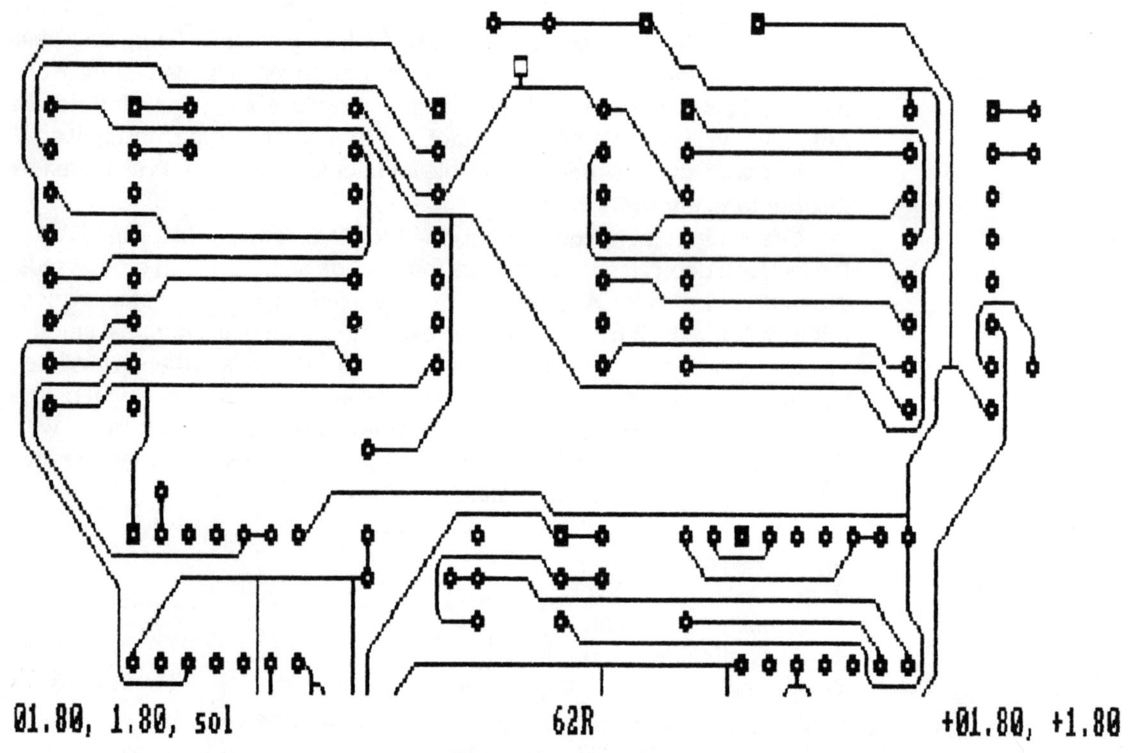

Fig. A-1. Placing a DIP pad pattern on the work area is a textual command operation.

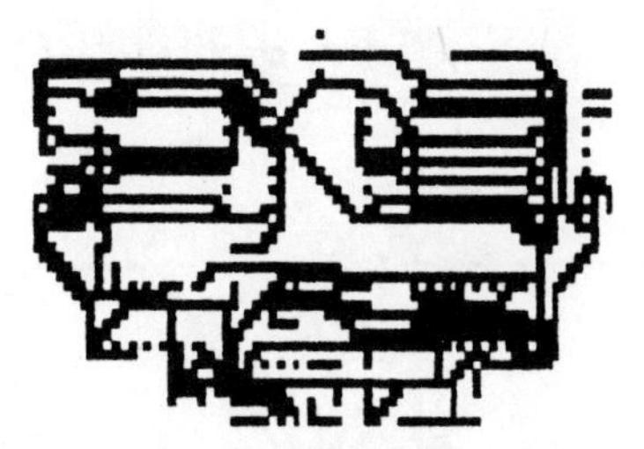

Fig. A-2. A window command reduces the overall "picture" of the current PCB for low-resolution orientation checks.

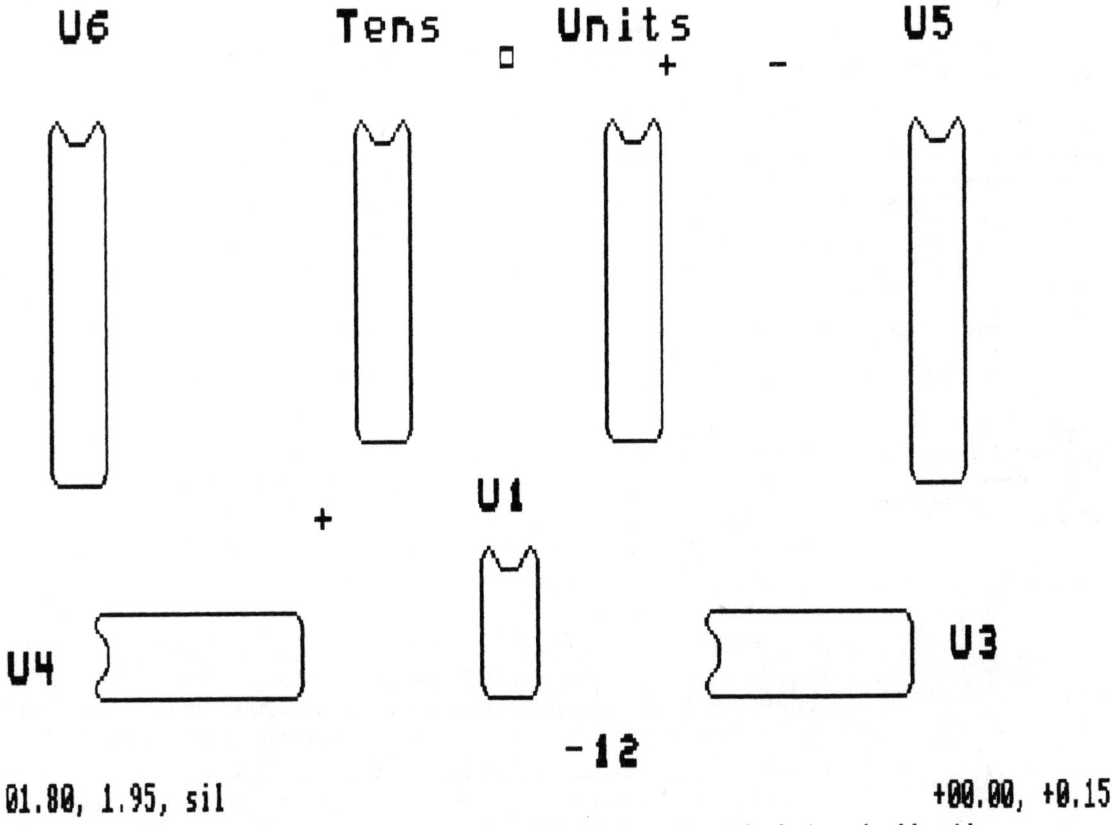

Fig. A-3. A silkscreen layer is generated simultaneously with placing of solder-side DIP pad patterns.

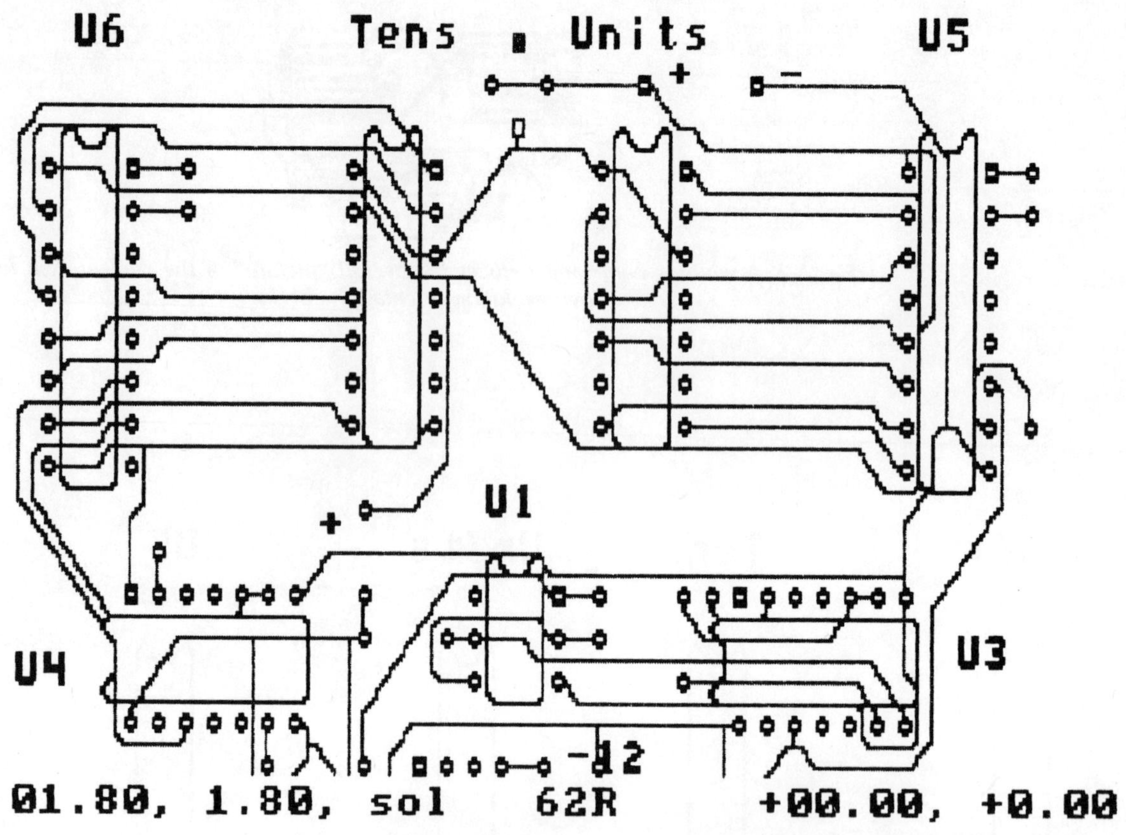

Fig. A-4. Both the solder layer and the silkscreen layer are continuously visible during layout when using an IBM Color Graphics Adapter.

introduction of the Apple Computer Macintosh. This graphics-based computer seemed ideally suited to the tasks of PCB CAD. Apparently, one different manufacturer shared this same opinion and created a breakthrough product for PCB design. QUIK CIRCUIT by Bishop Graphics is a full-featured CAD program that utilizes a graphics environment for preparing PCB layouts.

To provide a clearer picture of the operation of Quik Circuit, Figs. A-6 through A-10 illustrate the major steps involved in the design of a PCB.

A major point in the favor of these two PCB design programs is their low cost. Remarkably enough, you could purchase an entire Macintosh CAD system (i.e. 512K Mac, ImageWriter, and either Quik Circuit) for the same cost as AutoCAD 2. Now don't misinterpret this statement as a proclamation of equivalence between this Macintosh CAD package and either AutoCAD 2 or smARTWORK. The superiority of AutoCAD 2 and smARTWORK over this Macintosh program is clearly definable. This cost factor, however, is important to some designers and serves as a yardstick for measuring their

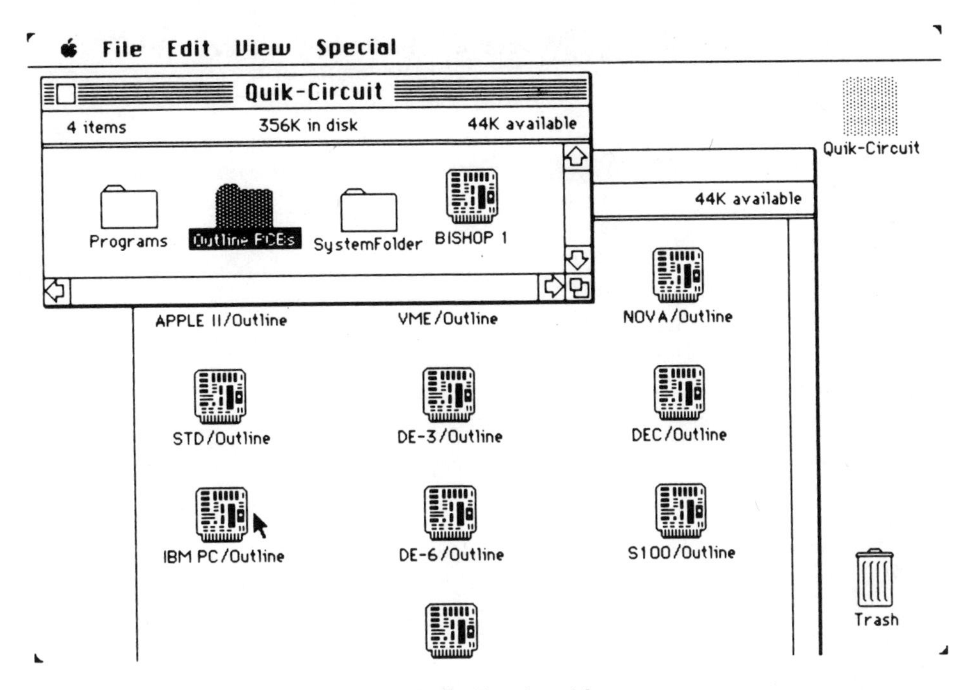

Fig. A-5. Quik Circuit's title screen.

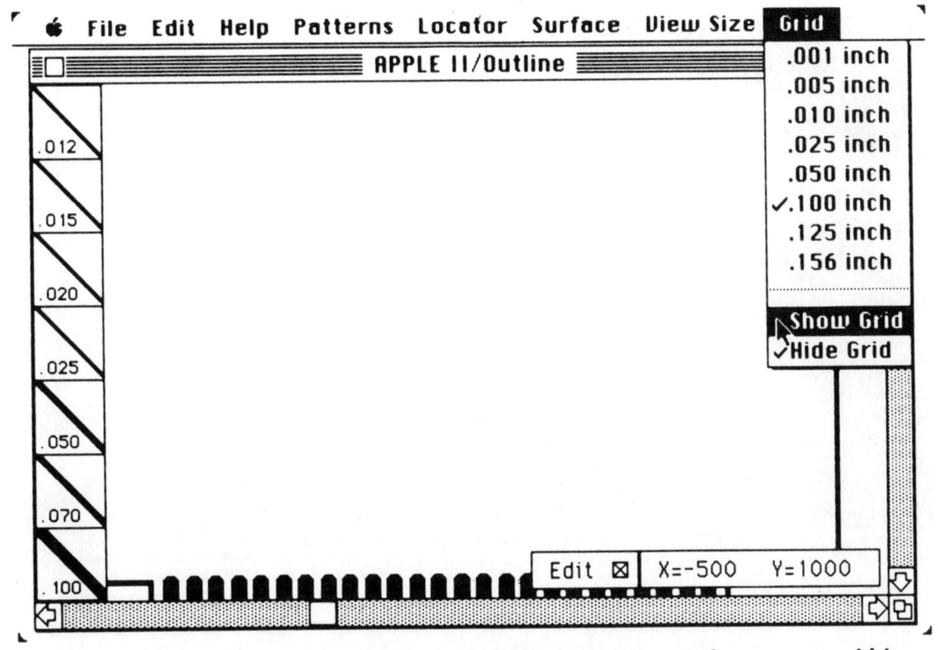

Fig. A-6. A temporary grid pattern can be placed over the work area as an aid in aligning pads and traces.

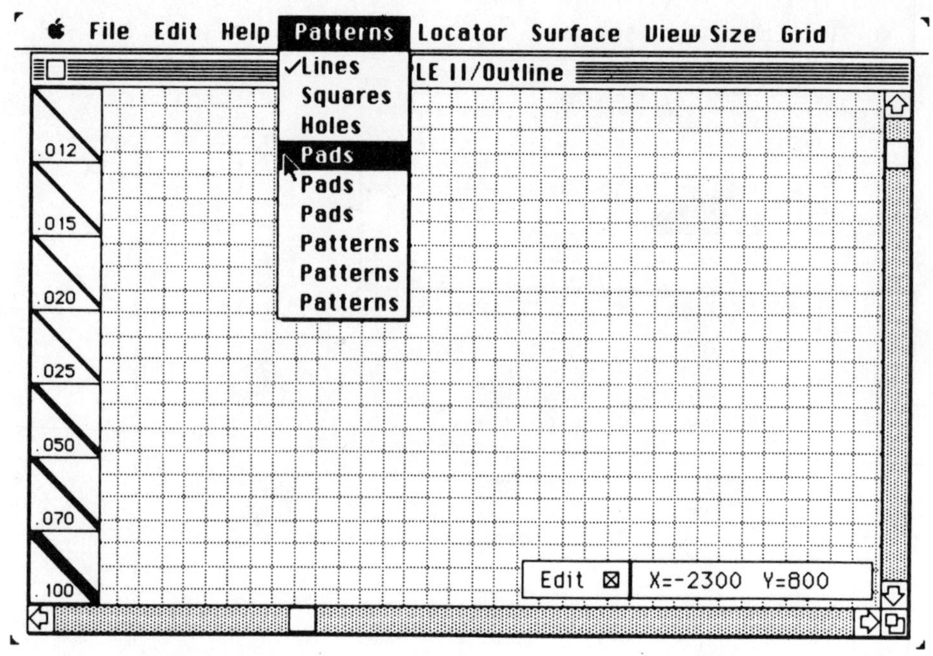

Fig. A-7. Positioning a DIP pad pattern on the work area involves the placement of individual pads through a selection from the Patterns menu.

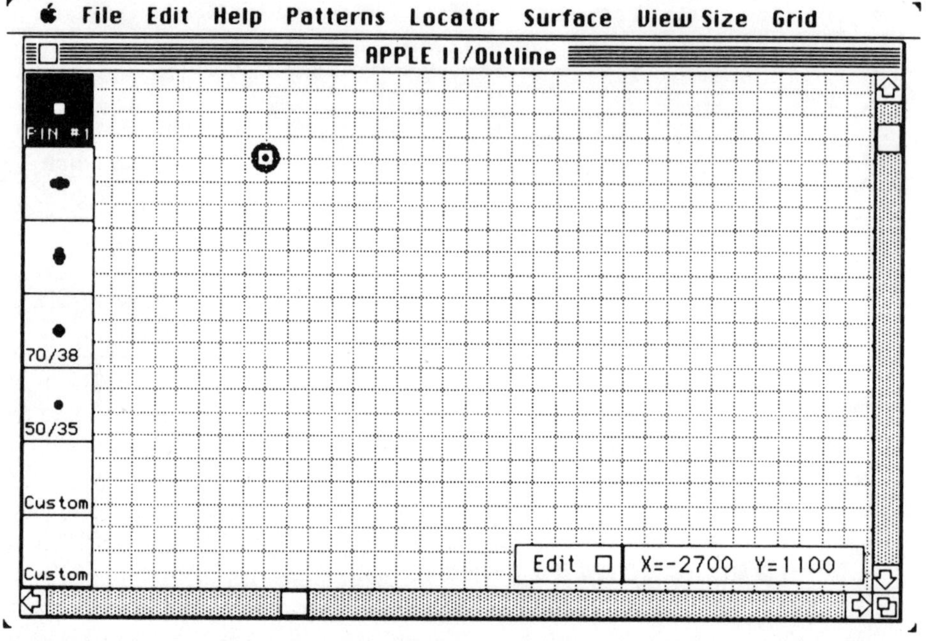

Fig. A-8. Pin #1 of the DIP is placed first, followed by the other pads of the DIP pattern.

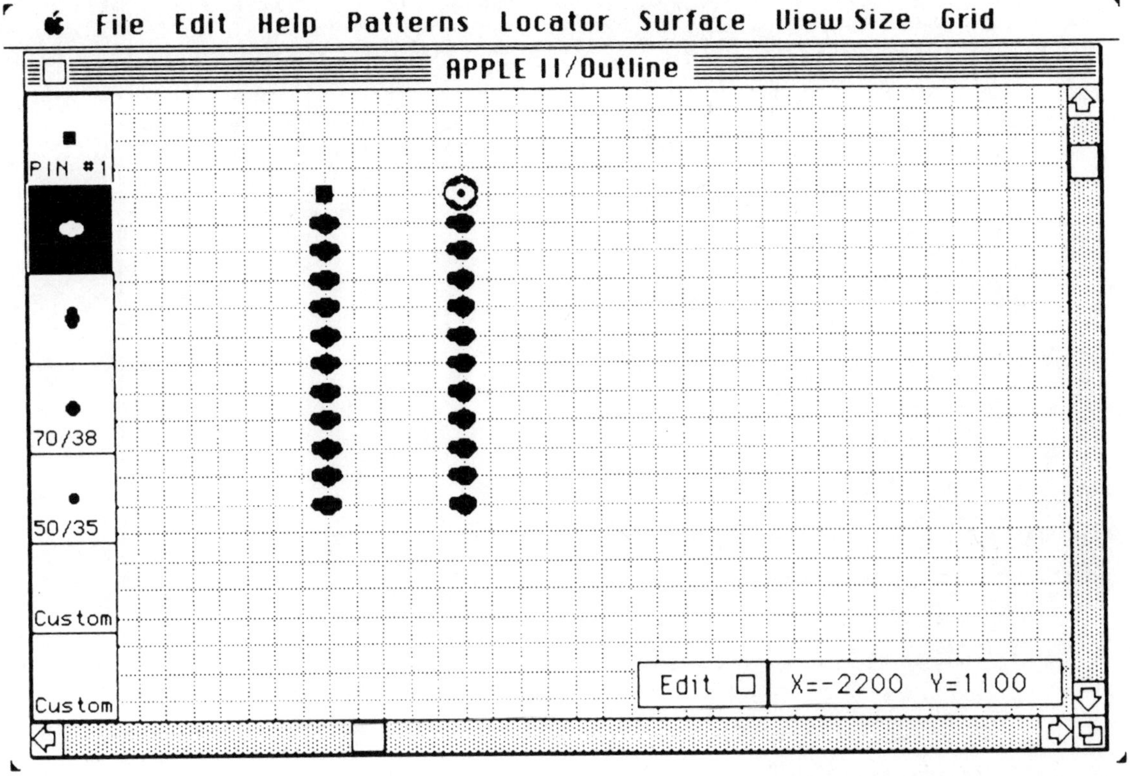

Fig. A-9. Following the establishment of pin #1, the other pads for the desired DIP pad pattern are placed on the work area.

solutions. A better and more easily defended argument that favors all PCB design software is with regard to the time saved over conventional circuit design methods.

ETCHING A PCB

Now that you know how a PCB template is made, your next step is learning how to make a PCB. But what are PCBs anyway? PCBs are vital for the mass production of circuit designs. The printed circuit board serves as the substratum for building any electronics circuit. The board itself is usually constructed from glass epoxy with a coating of copper on one or both of its sides. By using a powerful acid etchant like anhydrous ferric chloride, all of the copper that isn't protected by a resist (a substance that isn't effected by the action of the acid) is eaten away and removed from the PCB. This process leaves behind a copper tracing and/or pad where there should be an electrical connection. The leading method for placing the areas of resist on a PCB is with a photographic negative technique. Kepro Circuit Systems, Inc. markets

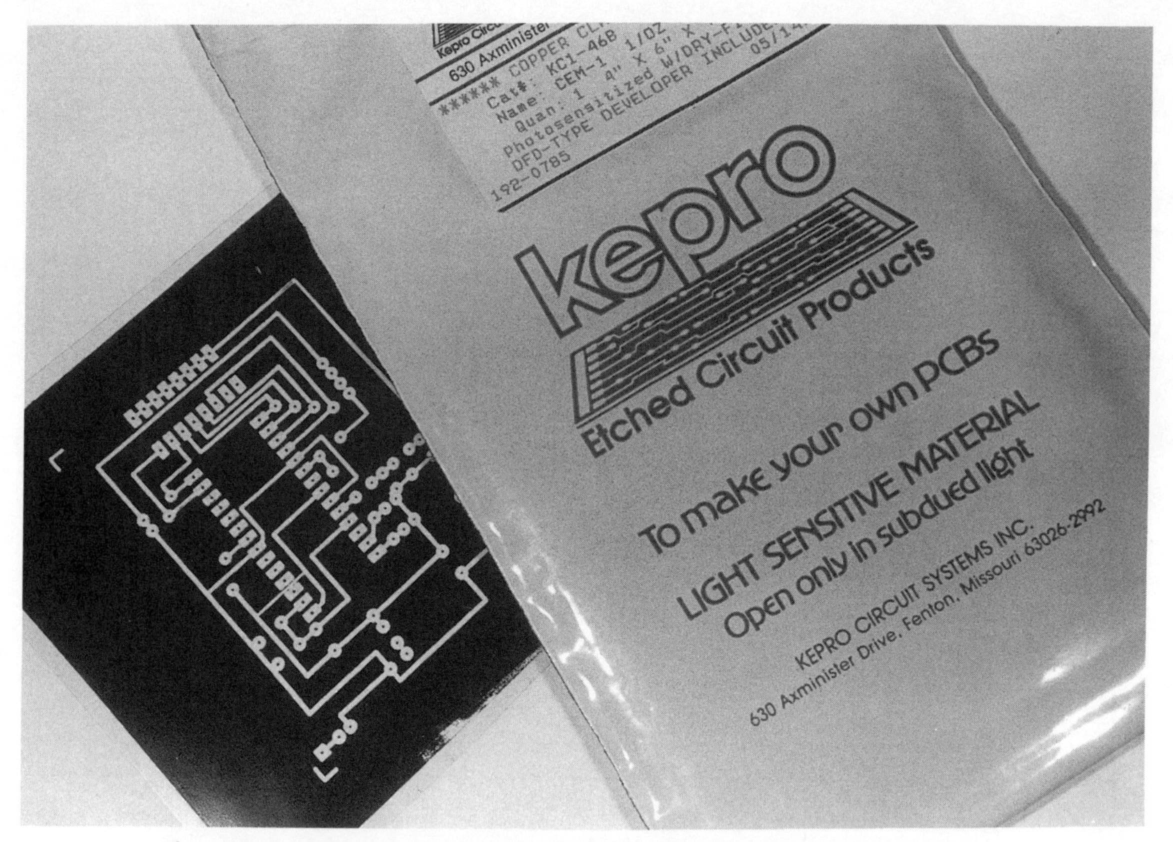

Fig. A-10. Kepro photosensitized circuit boards are ideal for preparing a working PCB.

a full range of pre-coated circuit boards for this purpose. These boards are excellent for PCB fabrication, and some, like the KeproClad Dry Film KC1-46B, come with their own developer and stripper (see Fig. A-10).

Basically, this technique acts exactly like its paper-based photographic cousin. In other words, parts of the PCB that are protected by the black portions of the negative are covered with a resist, while those regions that are exposed to the clear portions of the negative are removed by the acid.

Appendix B

IC Data Sheets

1458

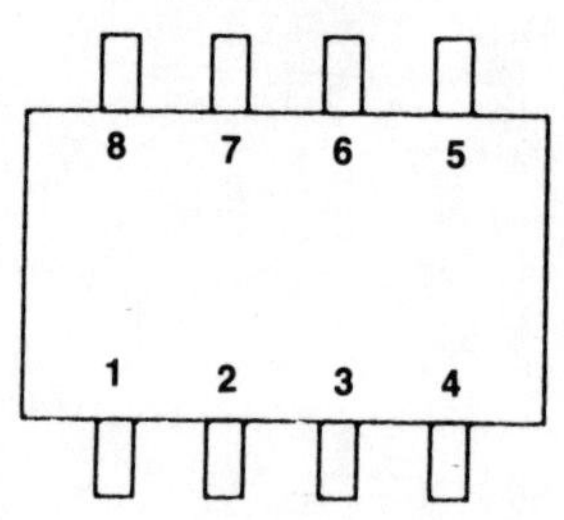

Pin Assignments	
Pin Number	Function
1	OUT A
2	Inverting IN A
3	Non-Inverting IN A
4	V –
5	Non-Inverting INB
6	Inverting IN B
7	OUT B
8	V +

3914

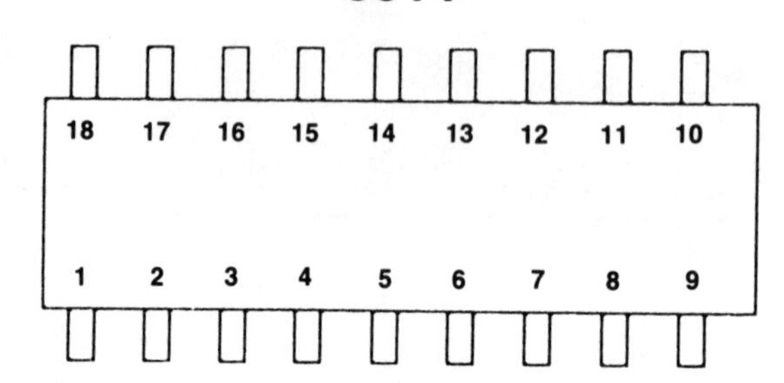

Pin Assignments			
Pin Number	Function	Pin Number	Function
1	LED1	10	LED10
2	V –	11	LED9
3	V +	12	LED8
4	Divider	13	LED7
5	Signal IN	14	LED6
6	Divider	15	LED5
7	Ref OUT	16	LED4
8	Ref Adjust	17	LED3
9	Mode Select	18	LED2

3916

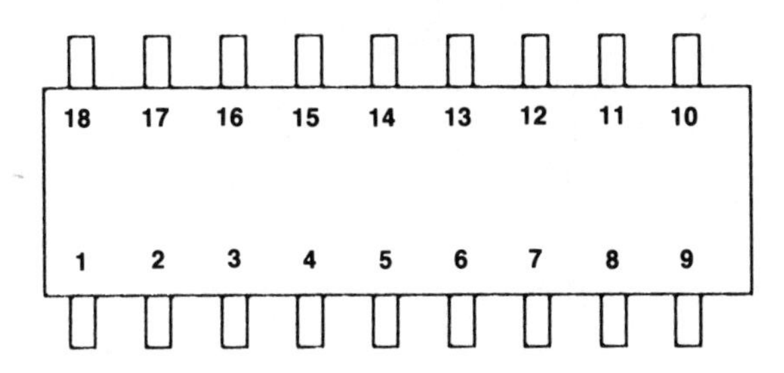

Pin Assignments			
Pin Number	**Function**	**Pin Number**	**Function**
1	LED1	10	LED
2	V−	11	LED
3	V+	12	LED
4	RL_o	13	LED
5	SIG	14	LED
6	RHi	15	LED
7	REF OUT	16	LED
8	REF ADJ	17	LED
9	Mode	18	LED

4013

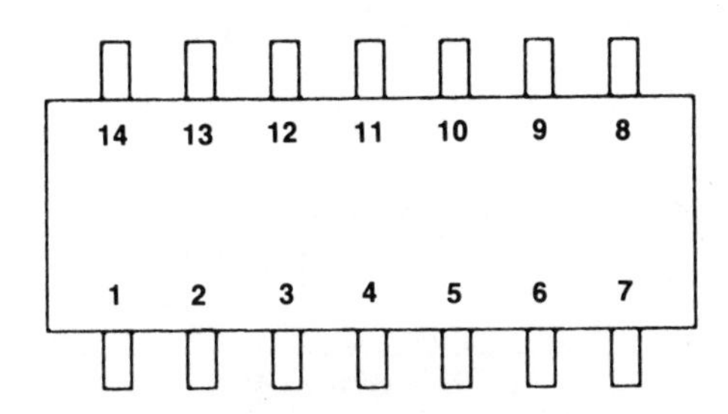

Pin Assignments			
Pin Number	**Function**	**Pin Number**	**Function**
1	Q1	8	Set2
2	$\overline{Q1}$	9	D2
3	Clock1	10	Reset2
4	Reset1	11	Clock2
5	D1	12	$\overline{Q2}$
6	Set1	13	Q2
7	V_{SS}	14	V_{DD}

555

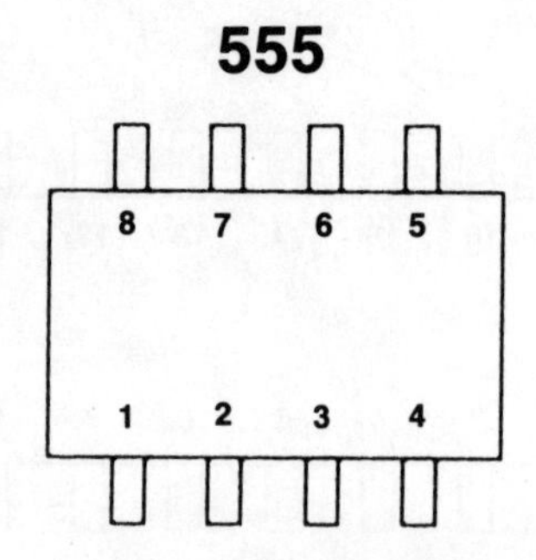

Pin Assignments	
Pin Number	Function
1	GND
2	Trigger
3	Output
4	Reset
5	Control Voltage
6	Threshold
7	Discharge
8	V_{CC}

741

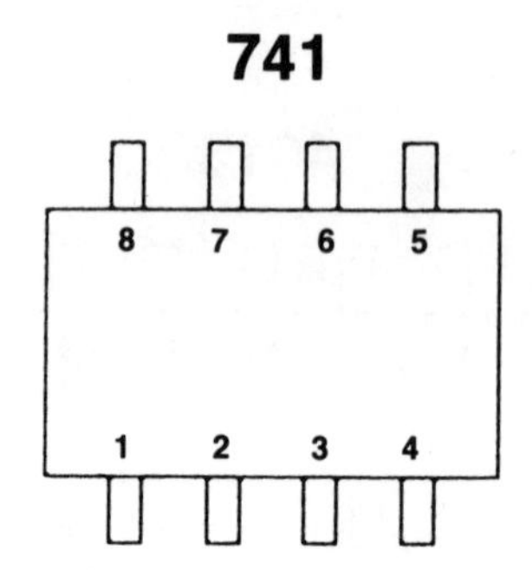

Pin Assignments	
Pin Number	Function
1	Offset Null
2	Inverting Input
3	Non-Inverting Input
4	V −
5	Offset Null
6	Output
7	V +
8	No Connection

74LS132

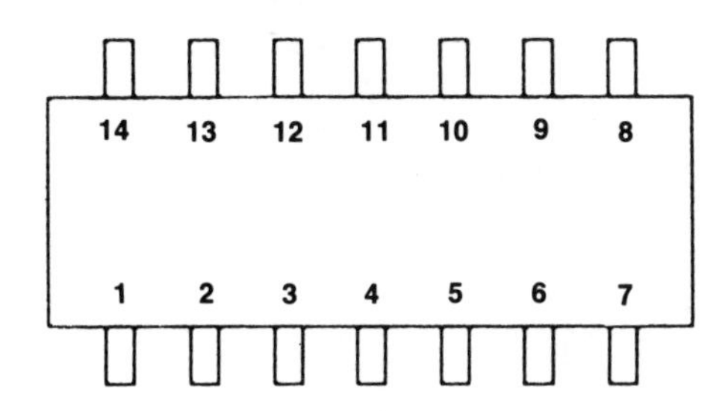

Pin Assignments			
Pin Number	**Function**	**Pin Number**	**Function**
1	1A	8	3Y
2	1B	9	3A
3	1Y	10	3B
4	2A	11	4Y
5	2B	12	4A
6	2Y	13	4B
7	GND	14	V_{CC}

74LS196

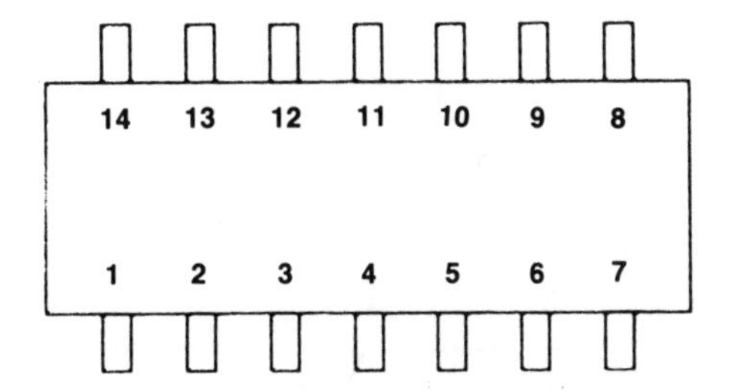

Pin Assignments			
Pin Number	**Function**	**Pin Number**	**Function**
1	Count	8	Clock 1
2	QC	9	QB
3	C IN	10	B IN
4	A IN	11	D IN
5	QA	12	QD
6	Clock 2	13	CLR
7	GND	14	V_{CC}

7447

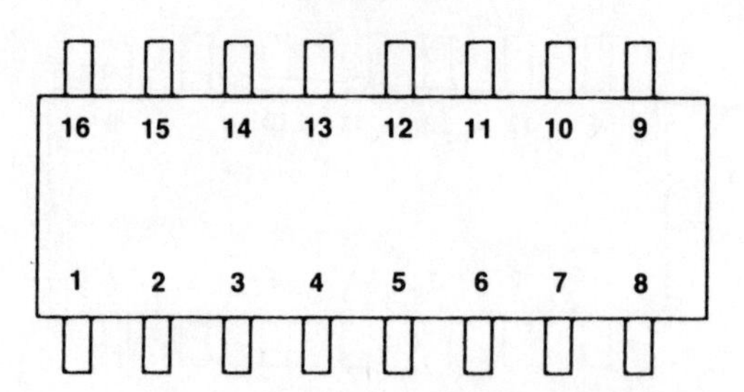

Pin Assignments

Pin Number	Function	Pin Number	Function
1	B IN	9	e OUT
2	C IN	10	d OUT
3	Lamp Test	11	c OUT
4	RB OUT	12	b OUT
5	RB IN	13	a OUT
6	D IN	14	g OUT
7	A IN	15	f OUT
8	GND	16	V_{CC}

74LS04

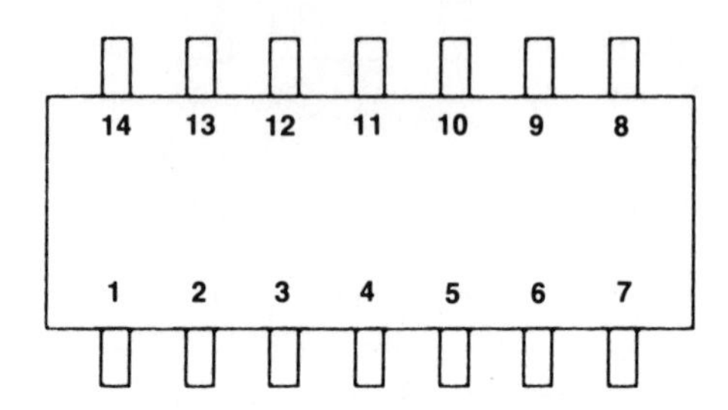

Pin Assignments

Pin Number	Function	Pin Number	Function
1	1A	8	4Y
2	2Y	9	4A
3	2A	10	5Y
4	2Y	11	5A
5	3A	12	6Y
6	3Y	13	6A
7	GND	14	V_{CC}

74LS08

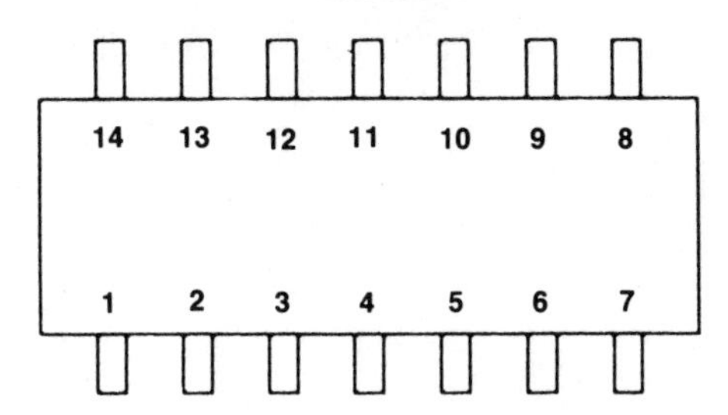

Pin Assignments			
Pin Number	Function	Pin Number	Function
1	1A	8	3Y
2	1B	9	3A
3	1Y	10	3B
4	2A	11	4Y
5	2B	12	4A
6	2Y	13	4B
7	GND	14	V_{CC}

74LS157

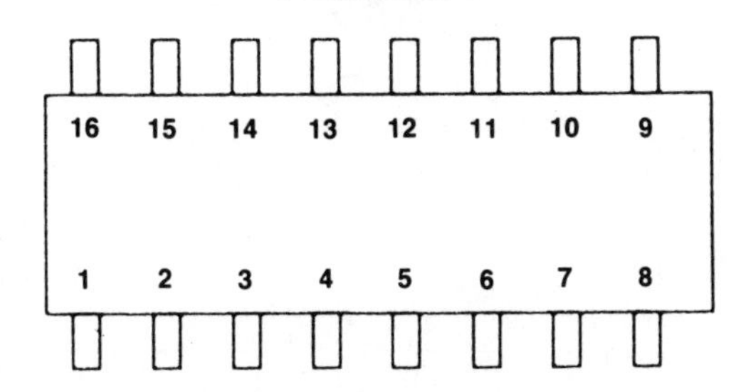

Pin Assignments			
Pin Number	Function	Pin Number	Function
1	SEL	9	3Y OUT
2	1A IN	10	3B IN
3	1B IN	11	3A IN
4	1Y OUT	12	4Y OUT
5	2A IN	13	4B IN
6	2B IN	14	4A IN
7	2Y OUT	15	Strobe
8	GND	16	V_{CC}

74LS30

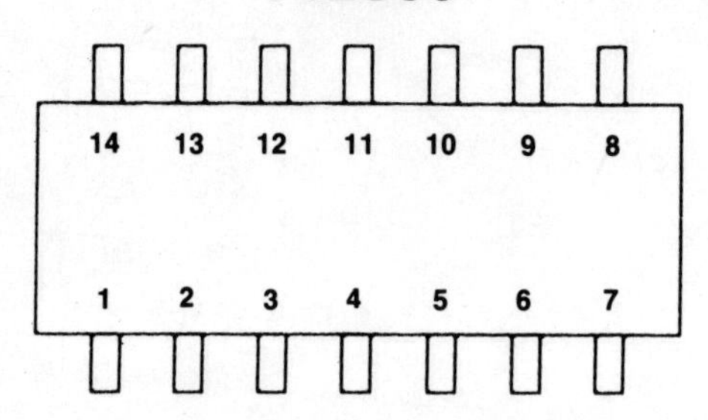

Pin Assignments			
Pin Number	Function	Pin Number	Function
1	1A	8	Y
2	B	9	NC
3	C	10	NC
4	D	11	G
5	E	12	H
6	F	13	NC
7	GND	14	V_{CC}

NE554

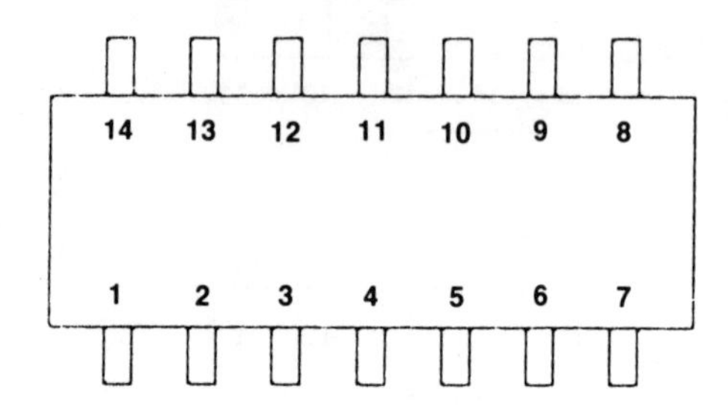

Pin Assignments			
Pin Number	Function	Pin Number	Function
1	Timing Cap.	8	Threshold
2	Timing Res.	9	A OUT
3	Reg. OUT	10	A PNP
4	IN	11	V +
5	GND	12	B PNP
6	Pulse Stretcher	13	B OUT
7	Deadband	14	Feedback

SN76832AN

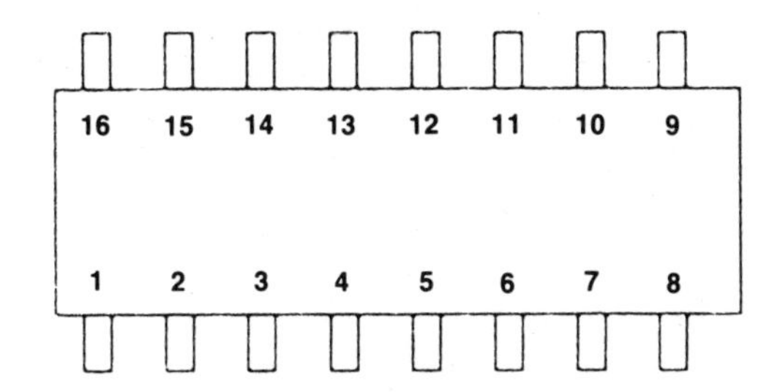

Pin Assignments			
Pin Number	Function	Pin Number	Function
1	VCO Timing Cap.	9	GND
2	VCO TIMING Res.	10	Amp Supp.
3	OUTPUT Enable	11	1st Amp
4	Data OUT	12	1st Amp
5	Loop Filter	13	1st Amp
6	2nd Amp Decoupl.	14	Lock Filt.
7	2nd Amp IN	15	GND
8	2nd Amp Decoupl.	16	V_{CC}

TL084CN

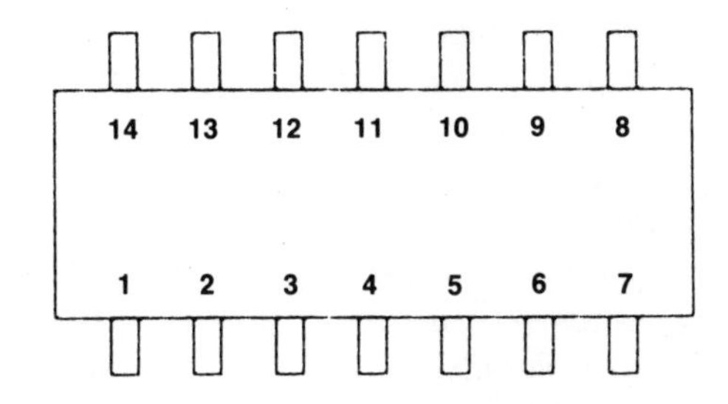

Pin Assignments			
Pin Number	Function	Pin Number	Function
1	OUT	8	OUT
2	Inverting IN	9	Invert IN
3	Non-Inverting IN	10	Non-In IN
4	$+V_{CC}$	11	$-V_{CC}$
5	Non-Inverting IN	12	Non-In IN
6	Inverting IN	13	Invert IN
7	OUT	14	OUT

Appendix C

Supply Source Guide

References to a number of unusual materials for constructing GaAs projects have been made throughout this book. Because some of these materials might be difficult to find in many remote areas, this appendix provides a list of mail order houses through which these items can be purchased. Additionally, the names and addresses of specific product manufacturers are included.

Apple Computer, Incorporated
20525 Mariani Avenue
Cupertino, CA 95014
ImageWriter
LaserWriter
Apple II family of computers (including the II+, IIe, and IIc)
Macintosh Computer

Autodesk, Incorporated
2658 Bridgeway
Sausalito, CA 94965
AutoCAD 2 software

Borland International
4585 Scotts Valley Drive
Scotts Valley, CA 95066
Turbo BASIC

Bishop Graphics, Incorporated
P. O. Box 5007
5388 Sterling Center Drive
Westlake Village, CA 91359
E-Z Circuit PC Boards (#EZ7402, #EZ7475, and #EZ7464)
E-Z Circuit Pressure-Sensitive Copper Patterns
Quik Circuit

Bytek Corporation
1021 South Rogers Circle
Boca Raton, FL 33431
System 125 PROM Programmer
WRITER-I
BUV-3 EPROM Eraser

CAD Software, Inc.
P.O. Box 1142
Littleton, MA 01460
PADS-PCB

Heath/Zenith
Benton Harbor, MI 49022
ET Trainer
ET-4200 Laser Trainer
ETA-4200 Laser Receiver

Intersil, Inc.
10600 Ridgeview Court
Cupertino, CA 95014
LM2907
ICM7226B
ICM7243A
ICL7107

Jameco Electronics
1355 Shoreway Road
Belmont, CA 94002
ICs
LEDs
EPROMs
EPROM Programmer & Eraser

Kepro Circuit Systems, Inc.
630 Axminister Drive
Fenton, MO 63026
Kepro Pre-sensitized Circuit Boards

Newark Electronics
500 North Pulaski Road
Chicago, IL 60624
ICs
LEDs
IREDs
EPROMs

Scott Electronics Supply Corporation
4895 F Street
Omaha, NE 68117
Ungar System 9000
Weller EC2000 Soldering Station

Radio Shack Stores
64K 2764 EPROM
MFOE71
MFOD72
R9-56
B1001R
3916
Modular Breadboard Socket

Sig Manufacturing Company, Incorporated
401 South Front Street
Montezuma, IA 50171
Aeroplastic ABS plastic sheeting
Clear Plastic Sheets
X-Acto Saw Blades (#234)

Tower Hobbies
P. O. Box 778
Champaign, IL 61820
Satellite City's Super "T" cyanoacrylate *adhesive*

VAMP, Inc.
6753 Selma Avenue
Los Angeles, CA 90028
McCAD P.C.B.

Vector Electronics Company, Inc.
12460 Gladstone Avenue
Sylmar, CA 91342
Vector Plugboards

Wahl Clipper Corporation
2902 Locust Street
Sterling, IL 61081
Isotip Soldering Irons (#7800, #7700, and #7240)

Wintek Corporation
1801 South Street
Lafayette, IN 47904-2993
smARTWORK
HiWIRE

For Further Reading

BOOKS

GaAs IC Symposium Technical Digest, 1982 and 1983, IEEE, Inc., Piscataway, NJ

A compilation of invited and presented technical papers dealing with several different aspects of GaAs IC technology.

IEEE Gallium Arsenide Integrated Circuit Symposium, 1984, IEEE, Inc., Piscataway, NJ

Reprints of the papers that were presented at the October, 1984 IEEE Electron Devices Society GaAs IC symposium.

Mastering the 8088 Microprocessor, 1984, DAO, L.V., TAB BOOKS Inc., Catalog #1888

A thorough examination of the 8088 MPU and its command set.

Interfacing & Digital Experiments with Your Apple, 1984, ENGELSHER, C. J., TAB BOOKS Inc., Catalog #1717

All of the elemental electronics you need to know for plugging a circuit into your Apple computer.

Lasers—The Light Fantastic, 1987, HALLMARK, C. L. and D. T. HORN, TAB BOOKS Inc., Catalog #2905

An introduction to GaAs laser diodes.

How to Use Special-Purpose ICs, 1986, HORN, D.T., TAB BOOKS Inc., Catalog #2625

An interesting assortment of component data sheets for numerous digital and linear chips.

30 Customized Microprocessor Projects, 1986, HORN, D.T., Catalog #2705

30 different Z80-based projects are described along with an EPROM programmer.

Build Your Own Working Fiberoptic, Infrared, and Laser Space-Age Projects, 1987, IANNINI, R. E., TAB BOOKS Inc., Catalog #2724

The title says it all.

101 Projects, Plans, and Ideas for the High-Tech Household, 1986, KNOTT, J. and D. PROCHNOW, TAB BOOKS Inc., Catalog #2642

Over 100 circuit designs and ideas featuring many LED and fiber-optic applications.

Microprocessors and Logic Design, 1980, KRUTZ, R. L., John Wiley & Sons, New York, NY

A functional introduction into MPU and memory interfacing theory.

Microprocessor Architecture and Programming, 1977, LEAHY, W. F., John Wiley & Sons, New York, NY

A beginning text on implementing digital microcomputer designs.

The Handbook of Microcomputer Interfacing, 1983, LEIBSON, S., TAB BOOKS Inc., Catalog #1501

An excellent introduction to the electronics of parallel and serial connections.

The Laser Experimenter's Handbook, 1979, MCALEESE, F. G., TAB BOOKS Inc., Catalog #1123

Contains GaAs laser diode construction ideas.

Troubleshooting and Repairing the New Personal Computers, 1987, MARGOLIS, A., TAB BOOKS Inc., Catalog #2862

This book's odd title doesn't adequately convey the wealth of information on general microcomputer circuit design that is contained inside.

Chip Talk: Projects in Speech Synthesis, 1987, PROCHNOW, D., TAB BOOKS Inc., Catalog #2812

The definitive source on digital speech synthesis theory and speech synthesizer design.

Experiments in CMOS Technology, 1988, PROCHNOW, D. and D.J. BRANNING, TAB BOOKS Inc., Catalog #3062

Explores the facets of complementary metal-oxide semiconductor (CMOS) devices plus additional introductory material on some of the basics of various semiconductor technologies and theory.

Mastering the 68000 Microprocessor, 1985, ROBINSON, P.R., TAB BOOKS Inc., Catalog #1886

Complete data on the structure and command set found in the 68000 family of MPUs.

Microprocessors and Programmed Logic, 1981, SHORT, K. L., Prentice-Hall, Inc., Englewood Cliffs, NJ

All of the theory of MPU and memory interfacing that you'll ever need.

Applications of GaAs MESFETs, 1983, SOARES, R., J. GRAFFEUIL, and J. OBREGON, Artech House, Dedham, MA

An excellent reference on GaAs MESFET technology.

101 Projects for the Z80, 1983, TEDESCHI, F. P. and R. COLON, TAB BOOKS Inc., Catalog #1491

Hardware and software projects geared for the SD-Z80 System.

Handbook of Semiconductor and Bubble Memories, 1982, TRIEBEL, W. A. and A. E. CHU, Prentice-Hall, Inc., Englewood Cliffs, NJ

An introduction into memory technology.

MAGAZINE ARTICLES

"GaAs ICs and MMICs, Technology Close-Up," *Microwaves & RF*, Mar 1986, p. 63.

BYLINSKY, "What's Sexier and Speedier Than Silicon," *Fortune*, 24 Jun 1985.

CIARCIA, "Build a Serial EPROM Programmer," *BYTE*, Feb 1985, p. 105.

CURRAN, "GaAs-on-Silicon Wafers are Set to Go Commercial," *Electronics*, 15 Oct 1987, p. 47.

EDEN, LIVINGSTON, and WELCH, "Integrated Circuits: The Case for Gallium Arsenide," *IEEE Spectrum*, Dec 1983, p. 30.

GOSCH, "A Test That Could Boost GaAs Yields," *Electronics*, 28 May 1987, p. 36.

GRAHAM and SANDO, "Pipelined Static RAM Endows Cache Memories with 1-ns Speed," *Electronic Design*, 27 Dec 1984.

IVERSON, "How Silicon is Going to Copy GaAs," *Electronics*, 25 Jun 1987, p. 38.

TENNY, "EPROM Eraser," *Modern Electronics*, May 1987, p. 40.

WALLER, "Commercial Quantities of LSI GaAs are Finally Here," *Electronics*, 17 Sep 1987, p. 48.

Glossary

access time—the delay time interval between the loading of a memory location and the latching of the stored data.
address—the location in memory where a given binary bit or word of information is stored.
allophone—two or more variants of the same phoneme.
alphanumeric—the set of alphabetic, numeric, and punctuation characters used for computer input.
analog/digital (A/D) conversion—a device that measures incoming voltages and outputs a corresponding digital number for each voltage.
ASCII—American Standard Code for Information Interchange.
assembly language—a low-level symbolic programming language that comes close to programming a computer in its internal machine language.

binary—the base-two number system where 1 and 0 represent the on and off states of a circuit.
bit—one binary digit.
byte—a group of eight bits.

CCD—charge-coupled *d*evice; a SAM with slow access times.
chip—an integrated circuit.
chip enable—a pin for activating the operations of a chip.
chip select—a pin for selecting the I/O ports of a chip.
CMOS—a complementary *m*etal-oxide *s*emiconductor IC that contains both p-channel and n-channel MOS transistors.

CPU—central *p*rocessing *u*nit; the major operations center of the computer where decisions and calculations are made.

data—information that the computer operates on.

data rate—the amount of data transmitted through a communications line per unit of time.

debug—to remove program errors, or bugs, from a program.

digital—a circuit that has only two states, on and off, which are usually represented by the binary number system.

disk—the magnetic media on which computer programs and data are stored.

DOS—*d*isk *o*perating *s*ystem; allows the use of general commands to manipulate the data stored on a disk.

EAROM—electrically *a*lterable *r*ead-*o*nly *m*emory; also known as Read Mostly Memory.

EEPROM—electrically erasable *p*rogrammable *r*ead-*o*nly *m*emory; both read and write operations can be executed in the host circuit.

EPROM—an erasable *p*rogrammable *r*ead-*o*nly *m*emory semiconductor that can be user-programmed.

Field-Programmable Logic Array—a logical combination of programmable AND/OR gates.

firmware—software instructions permanently stored within a computer using a read-only memory (ROM) device.

floppy disk—see disk.

flowchart—a diagram of the various steps to be taken by a computer in running a program.

hardware—the computer and its associated peripherals, as opposed to the software programs that the computer runs.

hexadecimal—a base-sixteen number system often used in programming in assembly language.

input—to send data into a computer.

input/output (I/O) devices—peripheral hardware devices that exchange information with a computer.

interface—a device that converts electronic signals to enable communications between two devices; also called a port.

IRED—an LED radiating photons at infrared wavelengths.

languages—the set of words and commands that are understood by the computer and used in writing a program.

laser—the generation of coherent light radiation through the oscillations of atoms.

LED—a gallium arsenide-based substrate that emits photons when electrically stimulated.

loop—a programming technique that allows a portion of a program to be repeated several times.

LSI—a layered semiconductor fabricated from approximately 10,000 discrete devices.

machine language—the internal, low-level language of the computer.
memory—an area within a computer reserved for storing data and programs that the computer can operate on.
microcomputer—a small computer, such as the IBM PC AT, that contains all of the instructions it needs to operate on a few internal integrated circuits.
mnemonic—an abbreviation or word that represents another word or phrase.
MOS—a metal-oxide semiconductor containing field-effect MOS transistors.

NMOS—an n-channel metal-oxide semiconductor with n-type source and drain diffusions in a p substrate.
nonvolatile—the ability of a memory to retain its data without a power source.

octal—a base eight number system often used in machine language programming.
opcode—an operation code signifying a particular task to be performed by the computer.

parallel port—a data communications channel that sends data out along several wires, so that entire bytes can be transmitted simultaneously, rather than by one single bit at a time.
peripheral—an external device that communicates with a computer, such as a printer, a modem, or a disk drive.
phoneme—the basic speech sound.
PLA—see field-programmable logic array.
PMOS—a p-channel metal-oxide semiconductor with p-type source and drain diffusions in an n substrate.
program—a set of instructions for the computer to perform.

RAM—*r*andom *a*ccess *m*emory; integrated circuits within the computer where data and programs can be stored and recalled. Data stored within RAM is lost when the computer's power is turned off.
ROM—*r*ead-*o*nly *m*emory; integrated circuits that permanently store data or programs. The information contained on a ROM chip cannot be changed and is not lost when the computer's power is turned off.
RS-232C—a standard form for serial computer interfaces.

serial communications—a method of data communication in which bits of information are sent consecutively through one wire.
software—a set of programmed instructions that the computer must execute.
statement—a single computer instruction.
static—a RAM whose data is retained over time without the need for refreshing.
subroutine—a small program routine contained within a larger program.

terminal—an input/output device that uses a keyboard and a video display.

volatile—the inability of a memory to retain its data without a power source.

word—a basic unit of computer memory usually expressed in terms of a byte.

Index

PARTS LIST

Edited by Lisa A. Doyle